AF385765

R. Gottlob
R. May†

Venous Valves

Morphology
Function
Radiology
Surgery

With an Electron-optical chapter
by S. Geleff

Springer-Verlag Wien New York

Rainer Gottlob, M. D.
Professor of Experimental Surgery
First Surgical University Clinic of Vienna, Austria
Private address: Kirchengasse 28, A-1070 Wien, Austria

Robert May †, M. D.
Professor of Phlebology, Innsbruck, Austria

© 1986 by Springer-Verlag/Wien
Softcover reprint of the hardcover 1st edition 1986

Product Liability: The publisher can give no guarantee for information about drug dosage and application thereof contained in this book. In every individual case the respective user must check its accuracy by consulting other pharmaceutical literature.

The use of registered names, trademarks, etc. in this publication does not imply, even in the absence of a specific statement, that such names are exempt from the relevant protective laws and regulations and therefore free for general use.

With 98 partly colored figures

Library of Congress Cataloging-in-Publication Data. Gottlob, R. (Rainer). Venous valves. Includes index. 1. Veins–Valves–Diseases. 2. Veins–Valves. 3. Veins–Valves–Surgery. I. May, Robert. II. Geleff, S. III. Title. [DNLM: 1. Vascular Diseases. 2. Veins. WG 600 G686v.] RC695.G67. 1986. 616.1'4. 85-26080.

ISBN-13: 978-3-7091-8829-3   e-ISBN-13: 978-3-7091-8827-9
DOI: 10.1007/978-3-7091-8827-9

*Dedicated to our dear wives in gratitude
for their understanding and patience*

Preface

Venous valves rank among the smallest and most delicate organs of the human and animal bodies – so why devote an entire book to them? We were induced to do so by several reasons. First of all we would point out the clinical significance of venous valves. In the pathogenesis of a number of widespread diseases, such as varicose veins or the post-thrombotic syndrome, venous valves are involved as the underlying cause or at least a factor contributory to the symptoms. According to *Taheri et al.* these venous diseases occur ten times more frequently than arterial obliterations. Incompetence of venous valves also plays a causal role in varicocele, the most frequent cause of male infertility.

But not only pathogenetic reasons induced us to write this book. In more recent times there has been a growing tendency to reconstruct functional valve disorders therapeutically; several surgical methods have been developed, which are critically reviewed in this book.

It was our aim to sum up existing knowledge with respect to structure and function of venous valves and to expand that knowledge by findings of our own. Examinations of semi-thin sections and unilayered en-face preparations have hardly been published so far, and systematic studies of the ultrastructure by electron-microscopy were not to be found in the literature. We are very grateful, therefore, to Dr. Silvana *Geleff* for having undertaken such a study upon our suggestion.

In the discussion of the function of venous valves we particularly dealt with the results of venous pressure measurement. The method of photoplethysmography is also described and evaluated.

Many questions concerning the pathologic function of venous valves are still unsolved or controversial. Our aim is not to propagate our own views but to review controversial opinions critically.

The authors feel highly indebted to "Schweizerische Gesellschaft für Phlebologie" for a generous contribution that has made the publication of this book possible.

The Authors

Contents

Part I. History

According to *Franklin* (1927) three authors may have first discovered venous valves. Franklin argues for the claim made on behalf of Gian Battista *Canano*. That claim was supported by *Morgagni, Haller* and, in this century, by *Streeter* (1925). *Canano* was a professor of anatomy at Ferrara; he often met Vesalius and reported to him that he had found valves in the azygous vein, in renal veins and in veins of the sacral region. *Canano* did not publish his findings himself; the publication was made by Amatus *Lusitanus* (who also lived in Ferrara) (1551). *Amatus,* however, erroneously assumed that the valves served the purpose of forming an obstacle to the bloodflow from the azygous vein into the caval vein. He also quoted *Canano* in that respect. The error made Amatus the object of much ridicule so that (according to *Franklin,* 1927) venous valve research suffered a setback of several years.

Another potential discoverer of venous valves may be seen in Charles *Estienne* (1545), who found what he called "apophyses membranarum" in liver veins. Charles *Estienne* already considered venous valves a means of preventing backflow of blood analogously to the function of cardiac valves.

As a third potential discoverer Jacobus *Sylvius* must be listed, who found valves in the region of the mouth of the azygous junction. *Sylvius* lived from 1478 to 1555. In his posthumously published work "Isagoge Anatomica" (1555) he describes valves at the mouth of the azygous as well as in other major veins (jugular vein, brachial vein, crural veins) and in the trunk of the caval vein, where the vessel leaves the liver (!).

It appears that these findings were forgotten so that one may justly consider *Hieronymus Fabricius of Aquapendente* the actual discoverer, who in his 24-page volume "De Venarum Ostiolis" (1603) published a correct description of venous valves. *Aquapendente* (1537–1619) was a teacher of anatomy at the University of Padua.

Characteristic drawings from *Aquapendente's* work are shown in Figs. 1 and 2.

It is true that *Galen* already had knowledge of cusp-shaped semilunar structures in the lumen of veins. Some anatomists men-

tioned valves already before *Aquapendente*. *Aquapendente's* work,
however, is so clear and fundamental a description that we may well
leave his predecessors to the mist of prehistory. *Aquapendente* clearly
realized the regulatory function of venous valves. Thinking still
within the concept of *Galen's* physiology of circulation, however, he
assumed that it was the function of venous valves to "moderate
excessive congestion of blood". Knowing that William *Harvey* was in
Padua shortly before the printing of the small book on valves, we may
safely assume that he knew his teacher's work*.

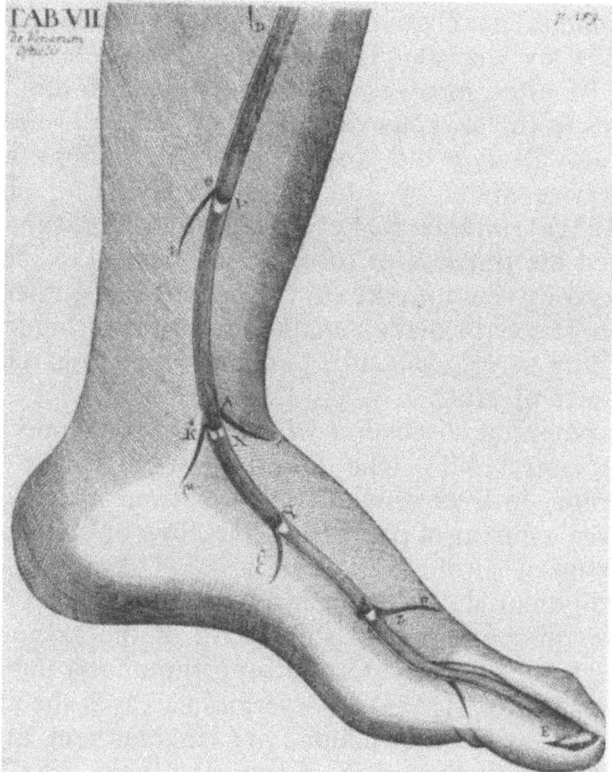

Fig. 1. Valves of the distal long saphenous vein (from Aquapendente)

It is interesting in this respect that, according to a note written
down in 1688 (i.e. after *Harvey's* death) by British scholar Robert

---

* We are grateful to Prof. Dr. *Hach,* Medical Director of William Harvey
Clinic, Bad Nauheim, and Prof. Dr. *Platzer,* Head of the Institute of
Anatomy at the University of Innsbruck, for having made available to us the
original books mentioned here.

Boyle (1627–1691), who was a friend of *Harvey's* during the years preceding his death, venous valves were of particular significance in the formation of his new concept of the circulatory system. Following *Galen's* concept, their structure was hard to explain indeed. William *Harvey,* born on April 1, 1578, in Folkstone at the Channel, published

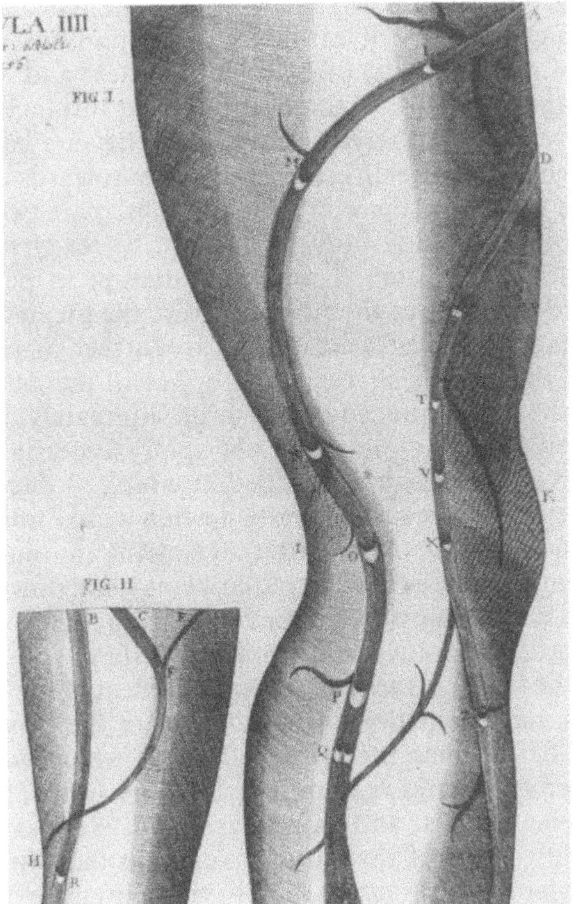

Fig. 2. Valves of a leg vein (from Aquapendente)

(Wilhelm Fitzer Press, Frankfurt am Main) a small book of 72 pages: "Exercitatio anatomica de motu cordis et sanguinis in animalibus" (Anatomic Treatise on the Motion of the Heart and the Blood in Animals).

This 72-page booklet contains our modern concept of the circulatory system and is comparable to no other work than Galilei's in physics. With it the Modern Age of Medicine began. The views *Galen* had held on the circulatory system were definitely revised. For that reason we would like to quote *Harvey* in his own words.

(Anonymous English translation, published by G. Moreton, 42 Burgate St., Canterbury, 1894.)

"The celebrated *Hieronymus Fabricius of Aquapendente,* a most skilful anatomist, and venerable old man, or, as the learned *Riolan* will have it, Jacobus Silvius, first gave representations of the valves in the veins, which consists of raised or loose portions of the inner membranes of these vessels, of extreme delicacy, and a sigmoid or semilunar shape. They are situated at different distances from each other, and diversely in different individuals; they are connate at the sides of the veins; they are directed upwards or towards the trunks of the veins; the two – for there are for the most part two together – regard each other, mutually touch, and are so ready to come into contact by their edges, that if anything attempt to pass from the trunks into the branches of the veins, or from the greater vessels into the less, they completely prevent it; they are further so arranged, that the horns of those that succeed are opposite the middle of the convexity of those that precede, and so on alternately.

The discoverer of these valves did not rightly understand their use, nor have succeeding anatomists added anything to our knowledge: for their office is by no means explained when we are told that it is to hinder the blood, by its weight, from all flowing into inferior parts; for the edges of the valves in the jugular veins hang downwards, and are so contrived that they prevent the blood from rising upwards; the valves, in a word, do not invariably look upwards, but always towards the trunks of the veins, invariably towards the seat of the heart. I, and indeed others, have sometimes found valves in the emulgent veins, and in those of the mesentery, the edges of which were directed towards the vena cava and vena portae. Let it be added that there are no valves in the arteries, and that dogs, oxen, etc. have invariably valves at the divisions of their crural veins, in the veins that meet towards the top of the os sacrum, and in those branches which come from the haunches, in which no such effect of gravity from the erect position was to be apprehended. Neither are there valves in the jugular veins for the purpose of guarding against apoplexy, as some have said; because in sleep the head is more apt to be influenced by the contents of the carotid arteries. Neither are the valves present, in order that the blood may be retained in the divarications or smaller trunks and minuter branches, and not be suffered to flow entirely into the more open and capacious channels; for they occur where there are no divarications: although it must be acknowledged that they are most frequent at the points where branches joint. Neither do they exist for the purpose of rendering the current of blood more slow from the centre of the body; for it seems likely that the blood would be disposed to flow with sufficient slowness of its own accord, as it would have to pass from larger into continually smaller vessels, being separated from the mass and fountain head, and attaining from warmer into colder places.

But the valves are solely made and instituted lest the blood should pass from the greater into the lesser veins, and either rupture them or cause them to become varicose; lest, instead of advancing from the extreme to the central parts of the body, the blood should rather proceed along the veins from the centre to the extremities; but the delicate valves, while they readily open in the right direction, entirely prevent all such contrary movement, being so situated and arranged, that if anything escapes, or is less perfectly obstructed by the cornua of the one above, the fluid passing, as it were, by the chinks between the cornua, it is immediately received on the convexity of the one beneath, which is placed transversely with reference to the former, and so is effectually hindered from getting any further.

And this I have frequently experienced in my dissections of the veins: if I attempted to pass a probe from the trunk of the veins into one of the smaller branches, whatever care I took I found it impossible to introduce it far any way, by reason of the valves; whilst, on the contrary, it was most easy to push it along in the opposite direction, from without inwards, or from the branches towards the trunks and roots. In many places two valves are so placed and fitted, that when raised they come exactly together in the middle of the vein, and are there united by the contact of their margins; and so accurate is the adaptation, that neither by the eye nor by any other means of examination, can the slightest chink along the line of contact be perceived. But if the probe be now introduced from the extreme towards the more central parts, the valves, like the floodgates of a river, give way, and are most readily pushed aside. The effect of this arrangement plainly is to prevent all movement of the blood from the heart and vena cava, whether it be upwards towards the head, or downwards towards the feet, or to either side towards the arms, not a drop can pass; all movement of the blood, beginning in the larger and tending towards the smaller veins, is opposed and resisted by them; whilst the movement that proceeds from the lesser to end in the larger branches is favoured, or, at all events, a free and open passage is left for it.

But that this truth may be made the more apparent, let an arm be tied up above the elbow as if for phlebotomy (A, A, Fig. 1). At intervals in the course of the veins, especially in labouring people and

those whose veins are large, certain knots or elevations (B, C, D, E, F)
will be perceived, and this not only at the places where a branch is
received (E, F), but also where none enters (C, D): these knots or
risings are all formed by valves, which thus show themselves
externally. And now if you press the blood from the space above one
of the valves, from H to O, (Fig. 2) and keep the point of a finger
upon the vein inferiorly, you will see no influx of blood from above;
the portion of the vein between the point of the finger and the valve O
will be obliterated; yet will the vessel continue sufficiently distended

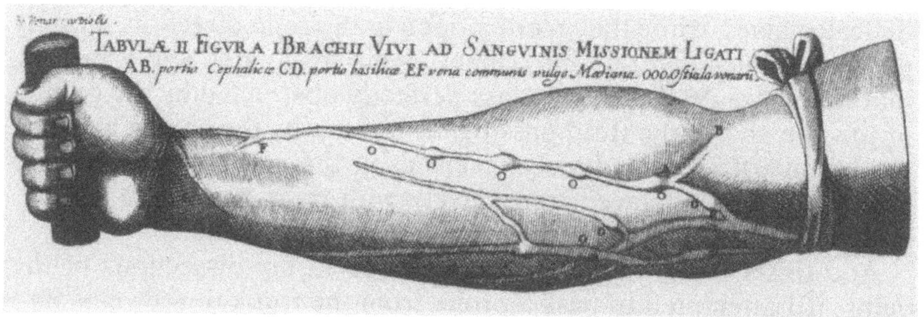

Fig. 3

Figs. 3 and 4. The only two illustrations contained in Harvey's "The Motion
of the Heart and the Blood". In the experiment during which the lower arm
is constricted, the venous valves are caused to protrude

above that valve (O, G). The blood being thus pressed out, and the
vein emptied, if you now apply a finger of the other hand upon the
distended part of the vein above the valve O (Fig. 3,) and press
downwards, you will find that you cannot force the blood through or
beyond the valve; but the greater effort you use, you will only see the
portion of vein that is between the finger and the valve become more
distended, that portion of the vein which is below the valve remaining
all the while empty (H, O, Fig. 3).

It would therefore appear that the function of the valves in the
veins is the same as that of the three sigmoid valves which we find at
the commencement of the aorta and pulmonary artery, *viz.,* to prevent
all reflux of the blood that is passing over them.

Further, the arm being bound as before, and the veins looking full
and distended, if you press at one part in the course of a vein with the
point of a finger (L, Fig. 4), and then with another finger streak the
blood upwards beyond the next valve (N), you will perceive that this

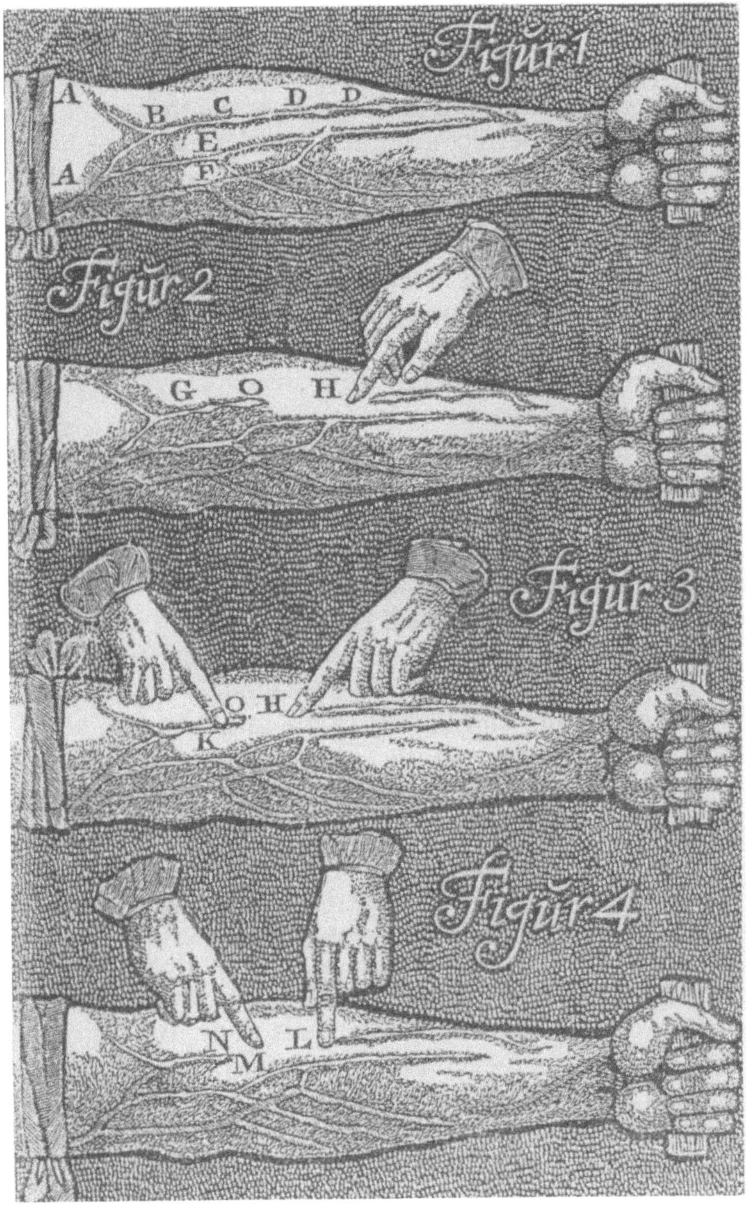

Fig. 4

portion of the vein continues empty (L, N) and that the blood cannot retrograde, precisely as we have already seen the case to be in Fig. 2; but the finger first applied (H, Fig. 2, L, Fig. 4), being removed, immediately the vein is filled from below, and the arm becomes as it appears at D, C, Fig. 1. That the blood in the veins therefore proceeds from inferior or more remote to superior parts and towards the heart, moving in these vessels in this and not in the contrary direction, appears most obviously. And although in some places the valves, by not acting with such perfect accuracy, or where there is but a single valve, do not seem totally to prevent the passage of the blood from the centre, still the greater number of them plainly do so; and then, where things appear contrived more negligently, this is compensated either by the more frequent occurrence or more perfect action of the succeeding valves, or in some other way: the veins, in short, as they are the free and open conduits of the blood returning *to* the heart, so are they effectually prevented from serving as its channels of distribution *from* the heart.

But this other circumstance has to be noted: The arm being bound, and the veins made turgid, and the valves prominent, as before, apply the thumb or finger over a vein in the situation of one of the valves in such a way as to compress it, and prevent any blood from passing upwards from the hand; then, with a finger of the other hand, streak the blood in the vein upwards till it has passed the next valve above (N, Fig. 4), the vessel now remains empty; but the finger at L being removed for an instant, the vein is immediately filled from below; apply the finger again, and having in the same manner streaked the blood upwards, again remove the finger below, and again the vessel becomes distended as before; and this repeat, say a thousand times, in a short space of time. And now compute the quantity of blood which you have thus pressed up beyond the valve, and then multiplying the assumed quantity by one thousand, you will find that so much blood has passed through a certain portion of the vessel; and I do now believe that you will find yourself convinced of the circulation of the blood, and of its rapid movement. But if in this experiment you will say that a violence is done to Nature, I do not doubt but that, if you proceed in the same way, only taking as great a length of vein as possible, and merely remark with rapidity at which the blood flows upwards, and fills the vessel from below, you will come to the same conclusion."

Summary Part I

By his life and work modern physiologic medicine based on experiment began. In particular, Harvey clarified the function of venous valves.

Part II. Healthy Venous Valves

1

Embryology of Venous Valves

The embryologic development of venous valves was investigated by *Jäger* (1926) as well as by *Kampmeier* and *La Fleur Birch* (1926–1927). In the jugular veins of pig embryos with head-to-coccyx sizes ranging from between 74 and 83 mm, *Jäger* found incipient bulgings of the lumen which constitute the preliminary stage of venous valves.

Kampmeier and *La Fleur Birch* studied human embryos. In the region of the sapheno-femoral junction they found the earliest traces of venous valves in embryos at 3.5 months. Towards the end of the fifth month a substantial number of venous valves could be found.

According to *Rickenbacher* (1966) the valves originate concurrently with the differentiation of the media in the endothelial tubes, which are devoid of any muscular layer up to this time.

Kampmeier and *Birch* divided the development of venous valves into five phases:

1. The first sign of an emerging venous valve is a *thickening of the endothelium,* which at this stage is largely inconspicuous and forms a pair of ridges placed transversely to the axis of the vessel. When these twin-ridges increase in size and touch each other, a ringlike structure evolves (Figs. 5 and 6).

2. Invaded by the mesenchyma situated below, the endothelial anlage grows; at the same time the vessel bulges out leeward of the valvular anlage.

3. The *evolving valve* directs itself toward the lee, the bulge on the upstream side of the valvular anlage increases, the outline becomes crescentic.

4. The valvular sac gains in *capacity,* the free edge of the valvular cusp *widens to nodular shape,* an effect which is caused by active involvement of the local mesenchyma.

5. The venous wall *thins down* considerably in the region of the valvular sinus; this process of thinning occurs mainly at the expense

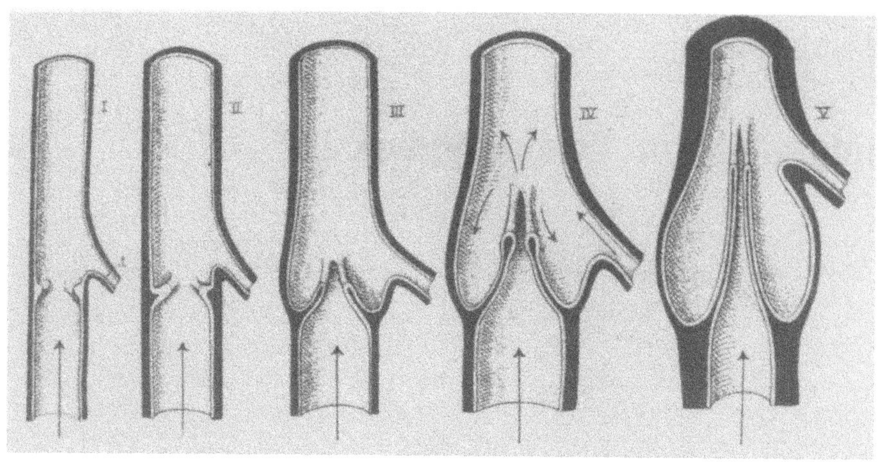

Fig. 5. Diagram of the development of a bicuspid valve. *I–IV* Phases I–IV as observed between the 3.5th and 5th fetal month. *V* state at full term. [From: Kampmeier and Birch, Amer. J. Anat. *38* (1936/37)]

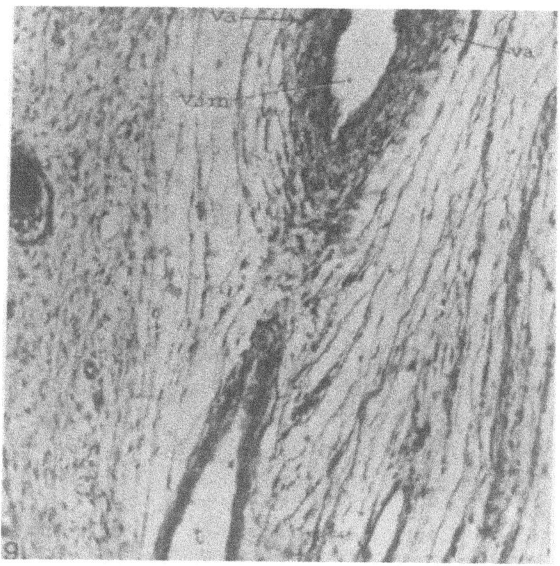

Fig. 6. Anlage of venous valve in internal saphenous vein *(vsm)*, phase I. *t* entering tributary. *va* valval anlage. [From: Kampmeier and Birch, Amer. J. Anat. *38* (1936/37)]

of the media, the remaining thickness of which is only one fifth of that observed in other regions of the vessel. In particular, the circular muscles become weaker. The *thickening* of the cusp margin observed in phase (4) *disappears;* a valve with clear-cut valvular function has come into existence (Fig. 5).

Jäger (1926) discusses the question of whether the development of the valvular anlage might be attributable to *physical factors.* It is to be assumed that certain turbulences will arise at the entrance of a small vein into a larger one or at the junction of two veins of equal diameter. *Jäger,* however, was unable to prove the existence of such turbulences through his model experiments. *Aschoff* (1912) ascribes a decisive role in the development of thromboses to these turbulences. In the presence of two obstacles projecting into the lumen in an oblique manner (Fig. 7) three eddies will form – one in front of the obstacle and two behind it.

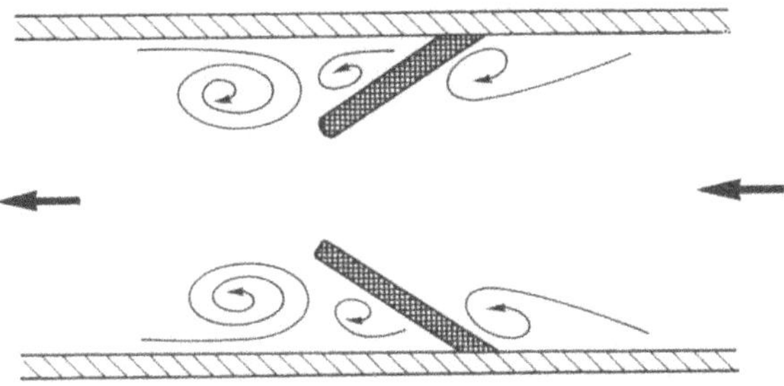

Fig. 7. Formation vortices in front of and behind oblique obstacles. L. *Aschoff* pointed out such turbulences as a possible cause of thrombosis. According to *Jäger* (1926)

However, it is to be emphasized that due to its smoothness and pliancy a sound venous valve will not be too capable of causing such turbulences. Instead, it will float passively in the bloodstream, yielding easily to all impulses of the current.

More recently, however, *Karino* and *Motomiya* (1982), studied the flow patterns in the pockets of venous valves, using isolated transparent dog saphenous veins and cinemicrographic techniques. Large paired vortices on both sides of the bisector plane of the valve leaflets were present in each valve pocket. Particles entered the valve pockets, described a series of spiral orbits of decreasing diameter and after long periods of time left the vortex, rejoining the main stream of the vein. A second, smaller, counter-rotating vortex was found deep in each valve pocket. The hematocrit in this secondary vortex was considerably lower than in the main stream of the vein.

2
Occurrence and Distribution of Venous Valves

A. General

Valves situated outside the venous system, such as cardiac or lymphatic valves, will not be dealt with in this book.

Still, we would not leave unmentioned here that there are also valve-like structures in arteries. *Wagenvoort* (1954) somewhat hesitatingly calls them "arterial aggers" but *Shanklin* and *Azam* (1963) refer to the presence of "valves" in cerebral arteries of rats. A histological comparison between arterial and venous valves was made by Böck (1975). Other than venous valves, arterial valves do contain smooth muscle cells, the basal membranes of which are in contact with the endothelial cells. In venous valves on the other hand we find an ample supply of collagenous fibrils, which – just as elastic material – have not been proved to be present in arterial valves.

In *animals* we find venous valves in amphibians, such as frogs (*Suchard,* 1907). In birds venous valves occur less frequently than in human beings; the distribution of venous valves in domestic animals is approximately equal to that in humans (*Ellenberger* and *Baum,* 1908).

In his classic novel *"Moby Dick"* *Melville* (1851) states that whales do not have venous valves. If injured, whales will lose blood extremely fast and in vast quantities.

Williams (1954) made a comparative study on the occurrence of valves in the veins of the extremities of monkeys, dogs and cats. Dogs were found to have the most venous valves, occurring more frequently in the superficial veins than in the deep ones.

According to *Franklin* (1927) valves do not occur in veins with a diameter of less than 1 mm. *Dzillas* (1949), however, describes valves in postcapillary veins with diameters from 20 to 145 μm. His findings are fully confirmed by *Spalteholz* and *Rulffs* (1958) as well as by other authors.

The *superior* and *inferior caval veins* are without valves. According to *Keith* (1907), these veins, along with the anonymous vein, the hepatic and renal veins, and the iliac veins, serve as a reservoir with a capacity of some 430 ml, which is of great importance for regulation of the circulatory homoiostasis.

The first venous valve cranial from the heart is found in the subclavian and internal jugular veins, not far from their confluence, which forms the anonymous vein. These valves have a certain importance in connection with infusion therapy. If the tip of an intravenous catheter is located on the cardiac side of these valves, the valves will close during the infusion, and the infused volume will reach the heart faster than if the application were made more cranially. For that reason *Schaeffer* (1973) called these veins *"border valves"* ("Grenzklappen").

Franklin (1969) describes a rudimentary unicuspid valve in the caval vein of humans, shortly below the entrance of a renal vein – undoubtedly a rarity.

In contrast to the clinical significance of leg and pelvic veins, the clinical significance of the valves in the branches of the superior caval vein is rather modest. For that reason we will mainly discuss those veins which drain the blood from the lower extremities. These valves are without doubt of the highest clinical importance. In addition thereto the valves of the internal spermatic veins are an important contributary factor of the development and treatment of male infertility (see pp. 122–127).

B. Leg and Pelvic Veins

The valves of leg and pelvic veins have been studied to great extent for their important role in connection with diseases of the legs caused by venous malfunction. Reports of varying comprehensiveness have been published by *Eger* and *Caspar* (1943), *Eger* and *Wagner* (1949), *Powell* and *Lynn* (1951), *Basmajian* (1952), *Lindvall* and *Lodin* (1961), *Ludbrook* (1968), *Reagan* and *Folse* (1971), *May* and *Nissl* (1973), *Staubesand* (1979), and *May* (1982).

a) Distribution of Valves in Various Leg Veins

We would emphasize some important aspects:

1. Frequency of occurrence of venous valves: Despite a great number of more recent works, the scheme established by *Ludbrook* (1968) has lost nothing of its validity.

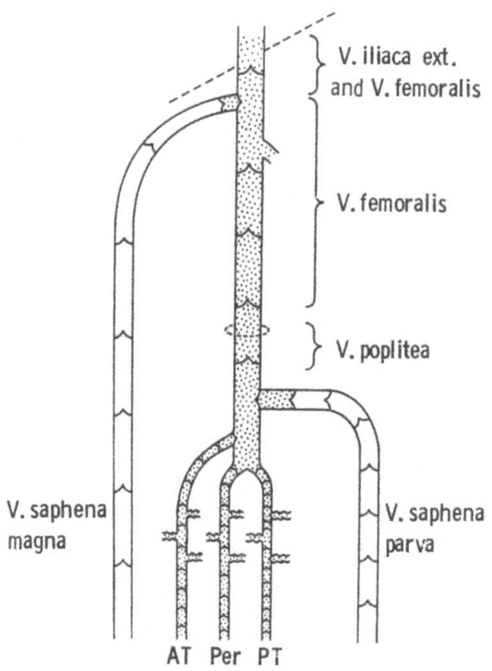

Fig. 8. Diagram of the venous system of the leg

Average Occurrence of Valves in Deep Veins

External iliac vein and common femoral vein (above sapheno-femoral junction) 1
(in 35 to 80%)
Femoral vein (below sapheno-femoral junction) 3
Popliteal vein 1
Posterior tibial vein 19
Anterior tibial vein 11
Peroneal vein 10

The valves of the gastrocnemic veins and the veins of the soleous muscle are subject to considerable variations. Typically, these veins are provided with an abundance of valves, which in the majority of cases will degenerate starting at age 30.

Superficial Veins

Long saphenous vein 7–9
Short saphenous vein 8
Communicating veins 2–3

2. The distance between valves in the deep veins is invariably shorter than in the superficial veins. In the superficial veins of the shank the distance is an average of 40 mm, in the deep veins 22 mm. This is important from a functional point of view, as the correct direction of the bloodflow from the surface into the deep venous system is thereby warranted.

3. The directional alignment of the valves in the perforating veins serves to ensure a unidirectional bloodflow from the superficial into the deep veins.

4. The perforating veins of the foot, in contrast, are either devoid of valves or provided with only one valve. This valve is aligned in such a way so as to allow the blood – other than as is the case in the leg – to flow physiologically, from the deep into the superficial veins (*Lofgren et al.,* 1968).

5. Number of valval cusps: As a rule, the valves are *bicuspid.* The highly important valve distal to the sapheno-femoral junction is often *tricuspid.*

6. The inferior caval vein is devoid of valves. Valves in the *common iliac vein* are rare findings (for a compilatory statistics consult *Lindvall* and *Lodin,* 1961). Above the sapheno-femoral junction valves are missing in 20 to 24% of cases (compilatory statistics according to *Reagan* and *Folse,* 1971).

7. Valves in superficial veins: In view of the fact that the most exhaustive work in this field is that of *Raivio* (1948), we will completely rely on his monography, neglecting a fair number of other authors.

8. Long saphenous vein: In the region of the thigh there are 3.5 valves on an average, in that of the lower leg 4 valves. In the region of the foot there are some 8 valves. The most important superficial valve is located slightly below the sapheno-femoral junction. According to *Cotton* (1961), the uppermost valve, located immediately at the entrance into the femoral vein, is an ostial valve. He refers to it as the "terminal valve". Some 5 cm further distal there is another valve, which *Cotton* refers to as the "subterminal valve" (Fig. 9, p. 26).

9. Short saphenous vein: 8.2 valves on an average.

10. Perforating veins: According to *Raivio* (1948) an average of 75% possesses valves. *Raivio* gives the figure 1–5 for the number of valves per perforating vein. Through comprehensive studies *Pirner* (1956) arrives at the following results: In all perforating veins dissected (obviously taken from phlebologically healthy subjects) there was evidence of valves, the maximum number being 3. If no more than one valve was found, is was located slightly before the junction with the deep system. Outside the muscular fascia, valves were invariably missing in perforating veins. No other than bicuspid valves were found. In many cases the perforating veins had branched out.

b) Lawful Regularity of Distribution ("Klappengesetze")

In 1880 *Bardeleben* laid down two so-called "laws" ("Klappengesetze") that have been quoted time and again in specialized literature.

1. A valve lies distal to every tributary and conversely there is a tributary-entry proximal to every valve.

2. The distance between valves invariably is a figure constant in each individual or a multiple thereof; the basic figure depends on the height, or rather limb-length, of the individual.

Bardeleben's assertions soon were vehemently contested by *Braune* (1889), *Klotz* (1887), *Friedrich* (1882), and *Meriel* (1926). Ever since the conclusive investigations by *Berutsen* (1925, 1927) and *Lanz* and *Wachsmuth* (1938) we may say that *Bardeleben's* "laws" have become untenable.

Bardeleben (1880) also maintained that the valves, whose occurrence was constant in fetal life, atrophied at a very early stage so that in 70-year-old adults merely 19% of the valves originally present were found. This was confirmed by *Klotz* (1887), *Hochstetter* (1887), *Lanz* and *Wachsmuth* (1938) and others. All these much-disputed questions may be said to have been conclusively answered by the comparatively more recent studies of *Raivio* (1948), *Powell* and *Lynn* (1951), *Jäger* (1926), *Kampmeier* and *Carrol* (1927), *Williams* (1954), *Saphir* and *Lev* (1952), *Luke* (1941), *Basmajian* (1952), and *Staubesand* and *Rulffs* (1958).

c) Do Venous Valves Perish During Post-fetal Life?

The assertion put forward by *Bardeleben* in 1880 stating that the venous valves developed during fetal life atrophied at a very early stage – so that no more than some 19% of the valves originally present would be found in a 70-year old adult – has become untenable in the form laid down. According to *Powell* and *Lynn* (1951) the reduction of valves is due to thromboses of the valvular sinus, whose course passes by unnoticed and whose occurrence is much more frequent than formerly supposed. On the other hand, *Leu et al.* (1979) have denied altogether that venous valves perish during post-fetal life and refused to regard venous valves as a primary factor in the development of varicosis.

These conflicting views may be brought to a common denominator if one compares the methods by which the authors arrived at their respective conclusions.

Klotz (1887) dissected internal saphenous veins from the distal side of the ankle to the junction with the femoral vein, examining the competence of proximal valves first by filling water into the vein in retrograde direction by means of a cannula. He evaluated the capability of valve-closing according to the respective degree of swelling of the vessel. Thereafter the vessel was cut open and the valve examined macroscopically. The procedure of function testing was then continued on the next valve in distal direction. The results of that combined method of examination are summed up in the following table (Table 1).

Table 1. *Results of Klotz's examinations (1887)*

A Corpse No.	B Age	C Number of functional valves	D Valves present by morphologic evidence	E Ratio between D and C in percent
VII	70.5	6	32	19
I	54	18	30	60
II	48.5	20	28	71
V	42	12	15	80
IV	30	22	26	85
VI	30	24	29	83
III	25	25	30	83
VIII	newborn inf.	36	36	100
X	newborn inf.	30	30	100
IX	32nd fetal week	35	35	100

Klotz's observations clearly show that the number of valves whose existence is demonstrable by morphologic evidence does not decrease, whereas with increasing age individual valves become incompetent. While almost a century later *Leu et al.* (1979) investigated valves of a different region of the venous system (external iliac vein, proximal segments of femoral vein and saphenous vein up to 6 cm below the sapheno-femoral junction) and arrived at the same morphologic findings, they did not give any information as to whether the valves they found were competent or not.

Based on the findings of *Klotz* as well as on observations of our own, we would advocate the following view with regard to those veins draining the blood from the lower extremities: Venous valves cannot just disappear in the course of a life – but pathologic processes may cause them to become incompetent and atrophied almost past recognition. Such pathologic processes may be:

1. Thrombosis of the valvular sinus. [Cf. *Paterson* and *McLachlin,* 1954, *Sevitt* (1974 a and b) and many others.]

2. Massive venous thrombosis. Valveless veins are a consequence of venous thrombosis and subsequent recanalization (*Homans,* 1916; *Edwards* and *Edwards,* 1937, and others). Thrombosis of the valvular sinus and sequelae of thrombosis will be discussed in more detail later.

3. Overdilation of the venous wall. It is obvious that, when the venous wall is extremely overdilated, the valves will be too short to touch each other and will become incompetent. There are indications to the effect that functional incompetence may even be followed by morphologic changes. *Edwards* (1934) observed that a collateral circulation via veins of the abdominal wall will form under thrombotic conditions of the iliac or caval veins. In the hypogastric region the collateral circulation will comprise the superficial epigastric veins and the circumflex iliac and pudendal veins, in which the bloodflow runs in orthograde direction, i.e. in the same direction as under normal conditions. In the epigastric region the collateral circulation is taken over by the internal mammary, intercostal and long thoracic veins. There the collateral circulation must have a direction of flow opposite to the normal direction. Under such circumstances the venous valves would be no more than an obstacle to the bloodflow. According to Edwards, these collaterals with retrograde bloodflow are valveless!

Several factors may be contributary to overdilation of the venous wall: increased hydrostatic pressure, as found with persons holding jobs requiring prolonged standing, or congestion caused by the gravid uterus. On the other hand, increased dilatability of vessels, as observed during pregnancy, may be of significance. *Szotér* and *Cronin* (1966) observed increased dilatability of lower-arm veins of patients with primary varicosis. It appears therefore that primary venous-wall weakness, as postulated also by *Leu et al.* (1979), does play an important role. – According to *Reagan* and *Folse* (1971) there seems to be a hereditary disposition, towards this weakness.

d) How Do Valves in Overdilated Veins Perish?

We would argue for the following sequence of events:

A vein is overdilated either because, as collateral vein of an obliterated main trunk, it suddenly has to transport an unphysiologically large quantity of blood, or because more proximal valves have perished, e. g. in the aftermath of valvular-sinus thrombosis, or because such valves have become incompetent due to dilation caused by abnormal load or weakness of wall. The vein is particularly subjected to dilation when the bloodflow changes to retrograde direction – as found in the abdominal-wall veins under conditions of thrombosis of the caval vein or in the long saphenous vein under conditions of varices. – Due to such dilation the venous valve itself becomes incompetent; blood regurgitates, and turbulences occur. Such turbulences, in turn, create favorable conditions for deposition of blood elements, which may be found on the valve itself or, even more frequently, in the valvular sinus. The ensuing processes of reorganization may result in shrinkage of the valve in that some portions of the valval cusp become attached to each other in folds or to the venous wall. Thus, in all forms of unphysiologic venous dilation, we find conditions ranging from morphologically unchanged but incompetent valves, or valves with micro-thrombi or sinus thrombi, to heavily shrunk or even completely missing valves. – The now incompetent valve will cause the valve located on the next site in distal direction to be subjected to exactly the same conditions of overdilation by excess load. In view of that fact, we consider the question of whether the morphologic change of the valve caused the venous dilation or vice versa to be as irrelevant and difficult to answer as the question of whether the hen preceded the egg or vice versa!

e) Veins in Other Circulatory Regions

So far we have been concerned with veins draining the blood from the lower extremities. We may assume with certainty that venous valves do perish spontaneously in other circulatory regions.

As early as in 1887 *Hochstetter* reported that the valves of the gastric veins become more and more incompetent with increasing age. At the age of twenty not a single competent venous valve was found in the major curvature.

In more recent times *von Kügelgen* and *Greinemann* (1958) investigated into renal veins. Their results may be summed up as follows: In $\frac{2}{5}$ of all kidneys examined, valves or rudiments of valves were found in the major renal vessels and their tributaries. Only a small portion of these valves were functional, and the majority of valves were found to be in various stages or regressive metamorphosis. In fetal and infant kidneys more and better preserved venous valves were present than in the kidneys of adults. These findings suggest the conclusion that processes of regression take place in the *ostial and free valves*. The situation is quite different in the *ostial valves* of the pelvic veins, which were present and obviously functional in all age groups.

These findings established by *Hochstetter* and *von Kügelgen* and *Greinemann* have well confirmed a working hypothesis first laid down by *Klotz* (1887) according to which the regressive changes of venous valves during post-fetal life are subject to regional differences. Venous systems, in which largely constant and unidirectional flow conditions are prevalent, tend to lose their superfluous valves more readily, whereas systems, in which heavier strain is exerted on the valves due to frequent arrest of flow or reversed flow, preserve their valves throughout life. (To which we would feel tempted to add: "unless, as will occur in some saphenous veins, such strain exceeds a certain limit of tolerance.")

3

Anatomy of Venous Valves

A. Macroscopic Anatomy

We distinguish between parietal valves and ostial valves.

a) Parietal Valves (Pocket Valves)

These valves are the type encountered most frequently. Most but not all of them are found at sites, where a tributary enters a vein further proximally or two veins of roughly equal diameter are united by a junction. (*Von Kügelgen* occasionally calls such valves "Astklappen", i. e. branch valves, a term which may be rather confusing as it is used by some other German-speaking authors for ostial valves and should therefore be avoided.) If a parietal valve is not in any relationship to an entrance or a junction it is also referred to be *free* parietal valve.

Parietal valves consist of the cusps, the valvular agger, the valvular cornua and the valvular sinus.

Valvular cusps are delicate – usually three-quarter to half-moon-sharped – structures, consisting of a skeleton of collagenous fibres covered by endothelium. Their insertion is at the valvular agger. Some portion of the collagenous fibres of the cusp runs in the form of a broad belt parallel to the free border of the cusp (*von Kügelgen,* 1958). "Occasionally a bundle of particular thickness is situated in the border of the cusp itself reinforcing the border in a beaded manner. In addition to the fibres running alongside the border of the cusp, bundles of fibres radiate from each of the cornua, flabellating across the cusp and re-entering the agger in the area between its medial line and the opposite horn. Thus the fibres of both sides cross and interweave one another in the central field of the valve."

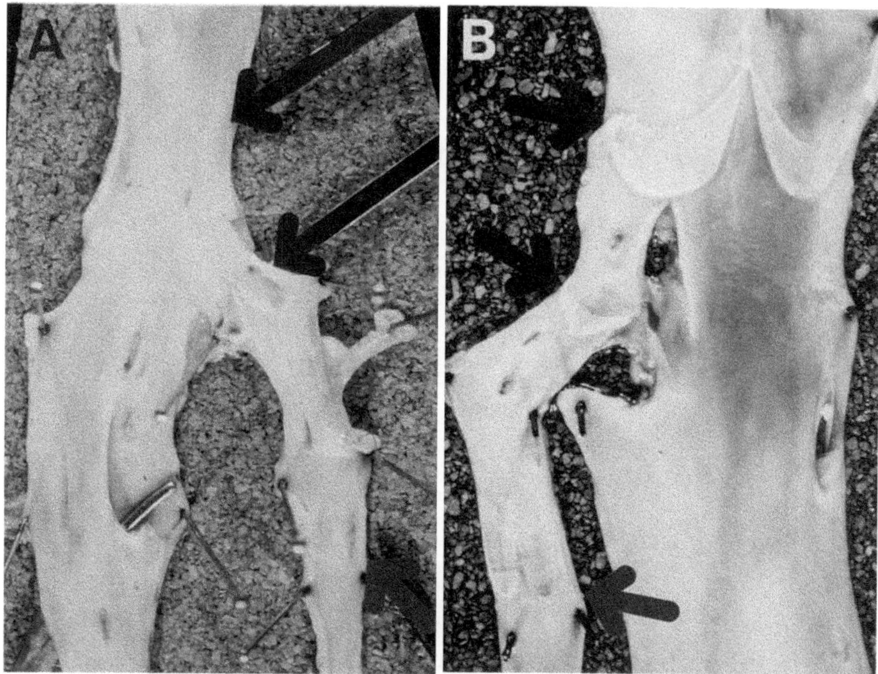

Fig. 9. Valves in the region of the sapheno-femoral junction. The left photograph **(A)** shows the typical conditions found. Upper arrow: valve in common femoral vein – visible but vaguely here. Middle arrow: terminal valve of long saphenous vein, immediately distally to the entrance. Lower arrow: subterminal saphenous valve. The probe has been introduced into the deep femoral vein. The photograph on the right-hand side shows somewhat untypical conditions. Upper arrow: valve in the common femoral vein, slightly proximal to the sapheno-femoral junction. Middle arrow: the terminal valve of the long saphenous vein is located at a distance of some 3 cm distally to the sapheno-femoral junction. Lower arrow: the subterminal valve of the saphena

Fig. 10. Diagram of bicuspid valve. **A** Longitudinal section of vein. **B** Opened vein seen from the lumen. *1* free border of cusp. *2* valvular sinus. *3* valvular agger. *4* valvular commissure, in which the cornua meet
Fig. 11. Valve of human femoral vein. Slightly proximal from the valve there is an entry

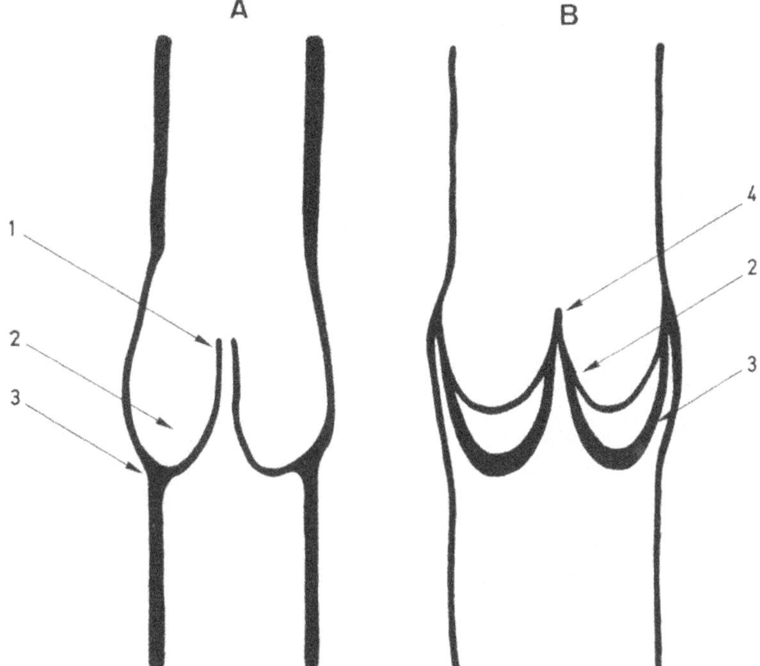

Fig. 10

Fig. 11

Valvular Agger

The valvular agger connects the valvular cusps to the venous wall; its shape is that of a double horseshoe, the convex sides of which are arranged distally. The valvular agger contains smooth muscle cells. It is not to be found in veins with diameters of less than 80 to 100 μm (*Staubesand* and *Rulffs,* 1958).

The *cornua* are formed at the juncture of the valvular aggers in the region of the commissure of the valve (Fig. 10). *Staubesand* and *Rulffs* (1958) refer to the cornua as the "dam of the commissure".

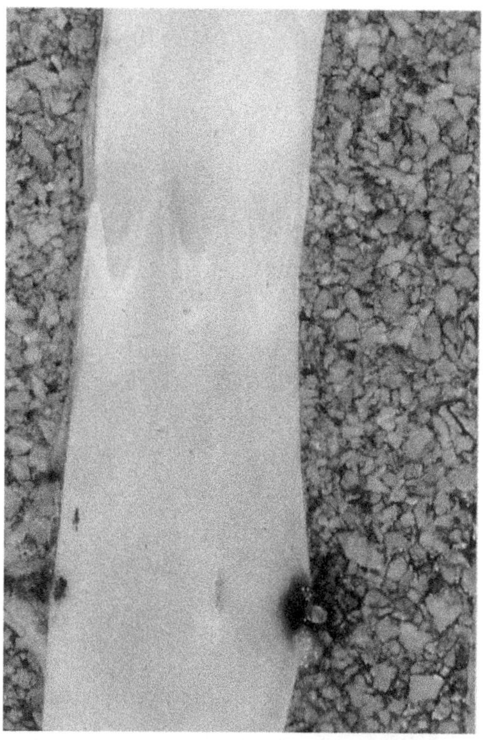

Fig. 12. Tricuspid valve in the jugular vein of a dog. (A rudimentary tricuspid valve found in a human saphenous vein is shown in Fig. 67)

The valvular pocket (sinus) is the space between valvular cusp and venous wall. In this region the venous wall is particularly thin.

Although parietal valves with 1 to 5 cusps have been reported, in human beings bicuspid valves are the rule. Still, tricuspid – as well as unicuspid – valves do occur occasionally. *Von Kügelgen* found quadricuspid valves in renal veins. *Von Kügelgen* (1958) argues that unicuspid valves may originate by regression of one cusp of a bicuspid valve.

In humans we have found a rudimentary tricuspid venous valve in only one instance (Fig. 67). Among numerous veins of rabbits and dogs examined, we found tricuspid valves in only one dog. In this particular dog, however, such valves were present in both femoral veins and in one long saphenous vein. Such multiple occurrence of tricuspid valves in one individual of the species (Fig. 12) would suggest a specific predisposition.

In a bicuspid valve the vertical length of the cusp will often be twice the diameter of the vessel (*Franklin*, 1927). According to *Edwards* (1934) the valvular cusps, when closed, touch each other over a length corresponding to $\frac{1}{5}$ to $\frac{1}{2}$ of the venous diameter at the respective site.

b) Ostial Valves

Ostial valves (occasionally termed "Astklappen" instead of "Mündungsklappen" in German specialized literature) are found immediately at the entry of a small vein into a larger vessel. According to *Franklin* (1927) ostial valves usually consist of a single fold, whose insertion occupies about two-thirds of the circumference of the entry. Ostial valves may also consist of two folds of unequal length, in which case oyster-shell-like structures may evolve.

According to *Franklin* ostial valves have no agger. *Von Kügelgen*, however, did find clear-cut valvular aggers in ostial valves (Fig. 13). He distinguishes between two different forms of ostial valves:

1. Marginal insertion of the ostial valve ("low" insertion according to *von Kügelgen*). The valve is seated directly at the circumference of the entry (Fig. 13, left). The distal agger is positioned entirely in the wall of the major vessel, the distal cusp freely hangs out into the lumen of the major vessel. There is no valvular sinus. The proximal cusp is provided with an agger situated at the site where the wall of the smaller vessel merges into that of the larger one. The sinus is located proximal from the agger of that cusp and completely in the wall of the major vein, which is considerably thinned in the area of the sinus.

A special variety of marginal ostial valves is the *funnel valve* (Fig. 14).

Franklin (1929), referring to a jugular vein, describes entries formed by spur-like structures, which are part of both the entering vein and the major vessel. This form of entry is obviously identical with the phenomena of marginal insertion and funnel valve.

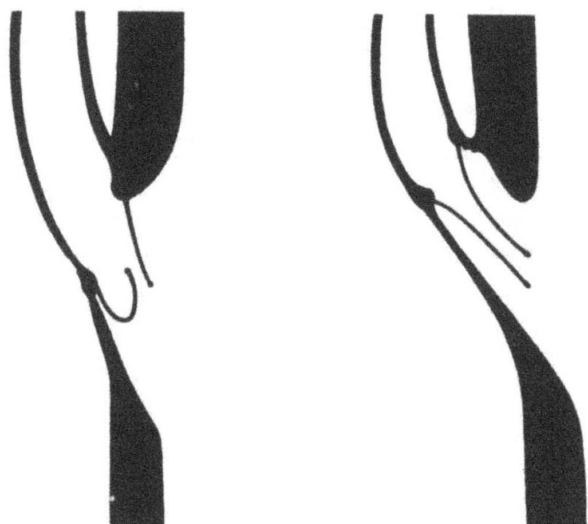

Fig. 13. Two types of ostial valves according to A. von Kügelgen [Zeitschr. Zellforsch. *47*, 684 (1958)]. Left: "low" insertion. Right: "high" insertion. The vessels shown are renal veins, in which the blood flows downward. In view of the fact that in many regions of the body the blood flows in cranial direction, we would recommend the use of the terms: "marginal" ostial valve (left) and "recessed" ostial valve (right)

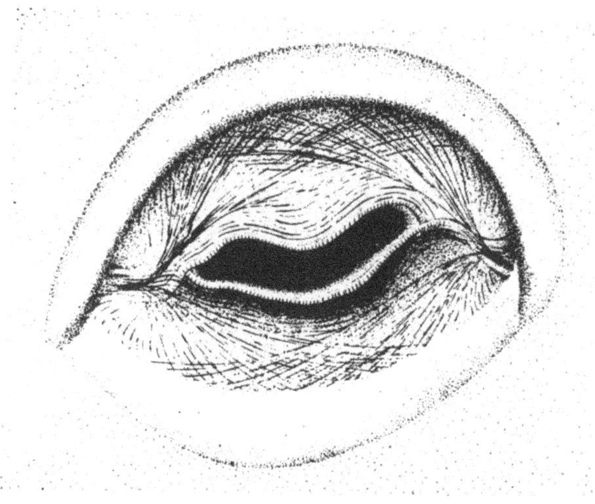

Fig. 14. "Funnel valve" according to A. von Kügelgen [Zeitschr. Zellforsch. *47*, 684 (1958)]

2. Recessed insertion of the ostial valve ("high" insertion according to *von Kügelgen*). The aggers of this type are located entirely in the entering vessel. Both halves of the valve have a sinus. In most cases the cusps hang out of the ostium and into the flow path of the major vessel (Fig. 13, right).

The nomenclature of "low" and "high" ostial valves as introduced by *von Kügelgen* is correct only if the blood flows downward in the vessel being referred to – as is the case in renal veins such as *von Kügelgen* studied. In many regions of the body, however, the blood must flow upward so as to reach the heart. There the nomenclature would be misleading. We, therefore, would recommend the use of the terms: ostial valve with "marginal" cusps and ostial valves with "recessed" cusps or simply "marginal" and "recessed" ostial valves.

Unicuspid ostial valves are largely identical with the spur that forms between the vessels if a small vein enters a larger one at an oblique angle. Presumably even slight hypertrophy of that spur will suffice to make it bend into the lumen of the entering vein if a retrograde pressure wave occurs. In this way the vein is protected against penetration by the pressure wave. Various forms of ostial valves are shown also in Figs. 27 and 28. Similar spur-like formations resembling valves are also found in the arterial systems, in the vicinity of entries (*Böck,* 1975).

c) Spatial Arrangement of Venous Valves

Fabricius (1603) stated that venous valves were arranged in such a manner that the next higher valve was always at right angles to the preceding one. *De l'Aulnoit* (1854) confirmed that theory. *Edwards* (1936) proved it to be erroneous and found out that venous valves are arranged according to entirely different regularities.

First of all *Edwards* observed that veins will assume cylindrical form only as specimens outside the body. Inside the body, veins will have cross sections of elliptic outline, in particular at the sites of venous valves. The elliptic outline is due to the fact that the skin, the subcutaneous fascia etc. press the vein towards the muscles or the muscles press it towards the bone (Fig. 15). As mentioned above, valvular cusps are aligned alongside the longitudinal axis of the ellipse and, thus, invariably parallel with the skin or the fascia surrounding the muscles. The correct arrangement of venous valves may be seen from Diagram B in Fig. 15. If the valve inserts into the venous wall at the longest diameter of the cross section, the cusps may join smoothly. If they were arranged in any other manner, e.g. in that shown in Diagram A of Fig. 15, the valve would form folds; close apposition between the valvular cusps would be impossible.

The spatial arrangement of valvular cusps may be important with respect to the angiographic method used for visualization of venous

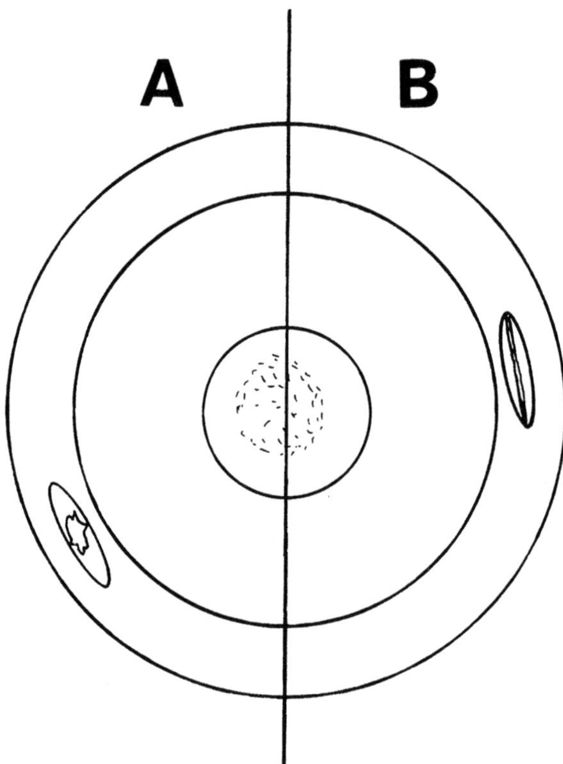

Fig. 15. Spatial arrangement of venous valves (according to Edwards, 1936). A The venous valve is arranged in correspondence with the shortest diameter of the ellipse. If that position were to occur in nature, valvular incompetence would result due to the excessive length of the cusps. B The valvular cusps insert in correspondence with the longest diameter of the ellipse – tight closing of the valve is possible. The outer circle represents the skin, the middle one the muscular fascia, and the inner one the outline of the bone

valves. In order to find out for purposes of subsequent surgery whether valves involved are competent, it may be desirable to obtain an orthograde image of their cusps. *Edward's* scheme shows in which way the extremity must be turned so as to obtain such an image.

Edward's findings were reaffirmed by *Samuels et al.* (1968). When trying to destroy venous valves by hooked stripper-type instruments, they realized that the edges of the valvular cusps were invariably aligned parallel to the skin. When the two hooks of the instrument were to be introduced into the valvular sinus, the instrument had to be turned so that the hooks were in radial position to the cross-section of the leg.

B. Histology

By light-optical means venous valves may be visualized in conventional sections (paraffin sections). A more refined method are semi-thin sections processed after embedding in synthetic resin (epon). Thirdly, there are "en-face preparations".

a) Paraffin Sections

1. Method

For the purpose of visualizing valves by means or conventional histologic methods, the vessel is removed along with the valve and fixed in a 5 to 10% formalin solution. Individual sections or series of sections in longitudinal or horizontal direction may now be made.

In order to fix the valves in closed position, we cannulized the veins proximally and distally to the valve. Neutralized formalin solution of 5 to 10% concentration was infused into the vein in orthograde direction; the vein was then suspended into a bottle containing formalin, and the central cannula connected to an infusion-set containing formalin. The upper level of the formalin solution was about 20 cm above the venous valve so that the valve was kept under a constant hydrostatic pressure effective in retrograde direction.

To fix the valvular cusps in a position as extended as possible, *Saphir* and *Lev* (1952) introduced into the valvular sinuses small cotton pads soaked with 10% formalin solution and immersed the entire vessel in formalin. – In view of the sensitivity of endothelia to mechanical trauma we would strongly advise against that procedure if the endothelial lining is to remain intact.

2. Results

The histologic picture as it presents itself in conventional sections is shown in Figs. 16 and 17. The valvular cusps may be divided into two regions, a luminal part facing the lumen of the vessel, and a parietal part facing the wall of the vessel. Both parts are covered by a layer of unicellular endothelium. Underneath the endothelium of the luminal part there is a thin elastic layer, which is an extension of the internal elastic membrane of the vein. The elastic membrane is slightly undulated. According to *Saphir* and *Lev* (1953) extremely thin elastic fibrils may branch off occasionally and invade the adjacent collagenous layer. The surface of the luminal part is relatively smooth, whereas the parietal part is rather irregularly outlined. Crypts and

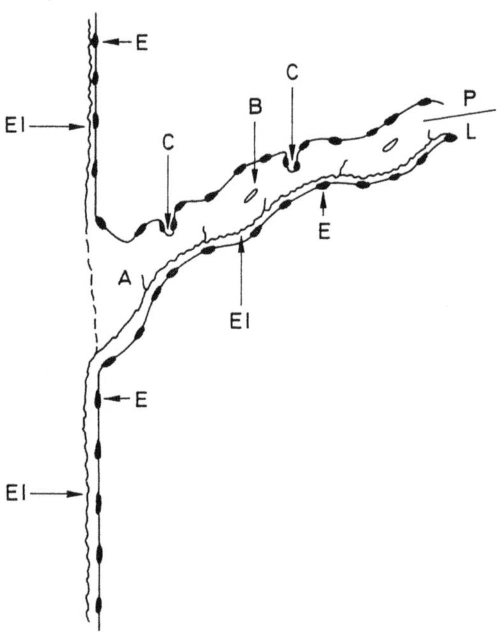

Fig. 16. Structure of a venous valve (scheme). *El* Elastic membrane. *E* Endothelial cells. *A* Valvular agger. *C* Crypts. *B* Connective tissue cell. *P* parietal part. *L* luminal part

crevices covered by endothelium invade the substance of the valves. These crypts are arranged in an irregular manner; occasionally they are absent. *Saphir* and *Lev* attributed the superficial irregularity of the parietal part to the absence of an elastic layer in that region. The stroma of the valve mainly consists of collagen; the valve contains few connective tissue cells. In the region of the valvular agger smooth muscle cells are found, but such cells do not invade farther into the valvular cusp.

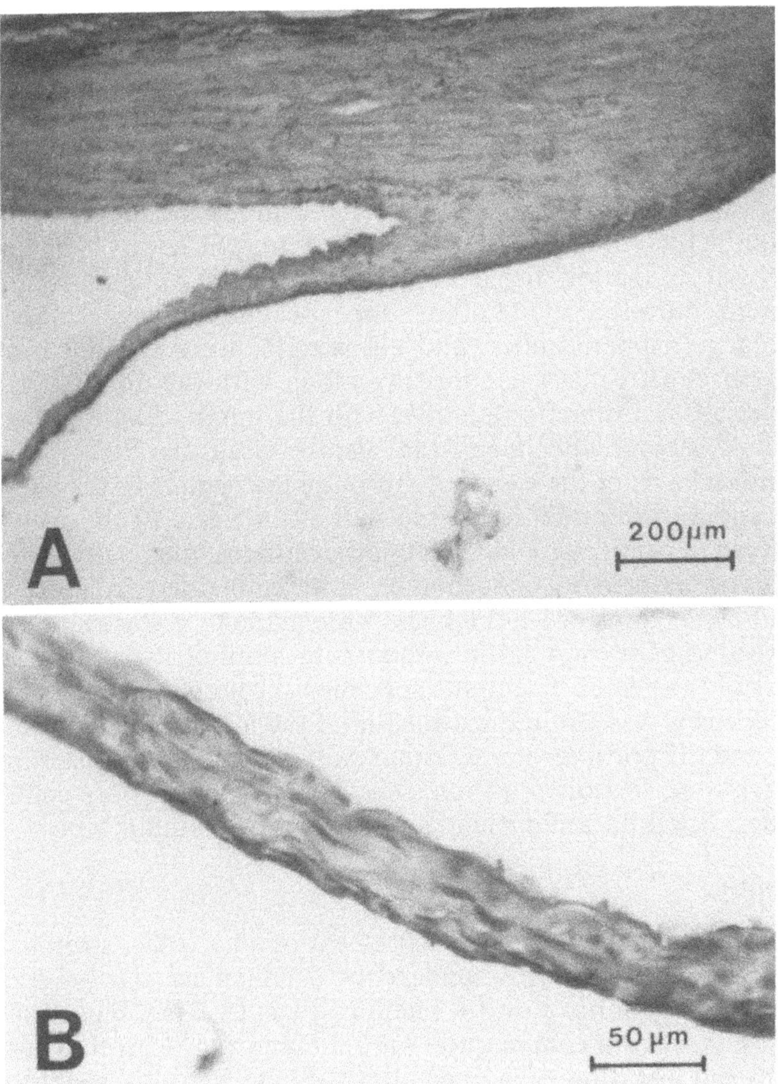

Fig. 17. Conventional paraffin sections of venous valves. **A** Insertion of the valve with valvular agger. Emergence of elastic membrane from venous wall into valvular cusp. Note the relatively smooth luminal part (bottom) and the crypts in the parietal part (top). – Femoral vein, 34-year old male. Orcein-hemalum staining. **B** Valvular cusp with endothelial and stroma cells, 18-month old infant, femoral vein. Hematoxiline-eosin staining

b) Semi-thin Sections

1. Method

The specimens were prepared as carefully and cautiously as the specimens for silver staining (see below). The veins were mounted on cork sheets, cut open, and spread out in their natural length and extension in such a manner that the intimal layer was facing upward. In that position the specimens were fixed in a 2.5% glutaraldehyde solution, which had been standardized to pH 7.2 by phosphate buffer. After fixation for a minimum duration of 2 hours and a maximum duration of 24 hours, the specimens were flushed by 0.2-molar phosphate buffer, and the samples were sectioned under the microscope. First a commissure along with the adjacent tissue was excised in a direction parallel with the longitudinal axis of the vessel. A second specimen was also excised parallel with the longitudinal axis of the vessel but through the centers of the valvular cusps, and a third one through the wall of the vessel. The specimens were fixed in osmium-pallade solution overnight, then dehydrated in alcohol of increasing concentration and embedded in epon 812. Thereafter, sections of 1 μ thickness were made by a glass knife, with the direction of section in the commissure running at cross angles to the axis of the vessel. The tissue specimen through the centers of the valvular cusps was cut in the direction of the longitudinal axis of the vessel, and the portions of the venous wall adjacent to the insertion of the valve were sectioned transversely. The specimens were collected on slides, dried on a hot plate, and stained by toluidin blue.

2. Results

In semi-thin sections details, such as endothelial cells, stroma cells and elastic lamellae, are visualized with enhanced clarity as compared to conventional paraffin sections (Figs. 18–21)*. Fig. 19 shows a section through a commissure. Fig. 21 A shows a semi-thin section through the insertion of a cusp; the valvular agger is visible. The valvular cusp itself is relatively smooth near the insertion of the parietal part. In contrast to that, Figs. 21 B and C show thick folds of the parietal part near the insertion of the cusp, i.e. at about the lowest point of the valvular sinus. Between the folds there are crypt-like indentations. Despite their vicinity to the valvular agger, they are not

---

* The semi-thin sections were prepared in cooperation with Prof. Dr. *Böck* and Dr. S. *Geleff,* both of the Institute of Micromorphology and Electron Microscopy of the University of Vienna. We feel greatly indebted to them.

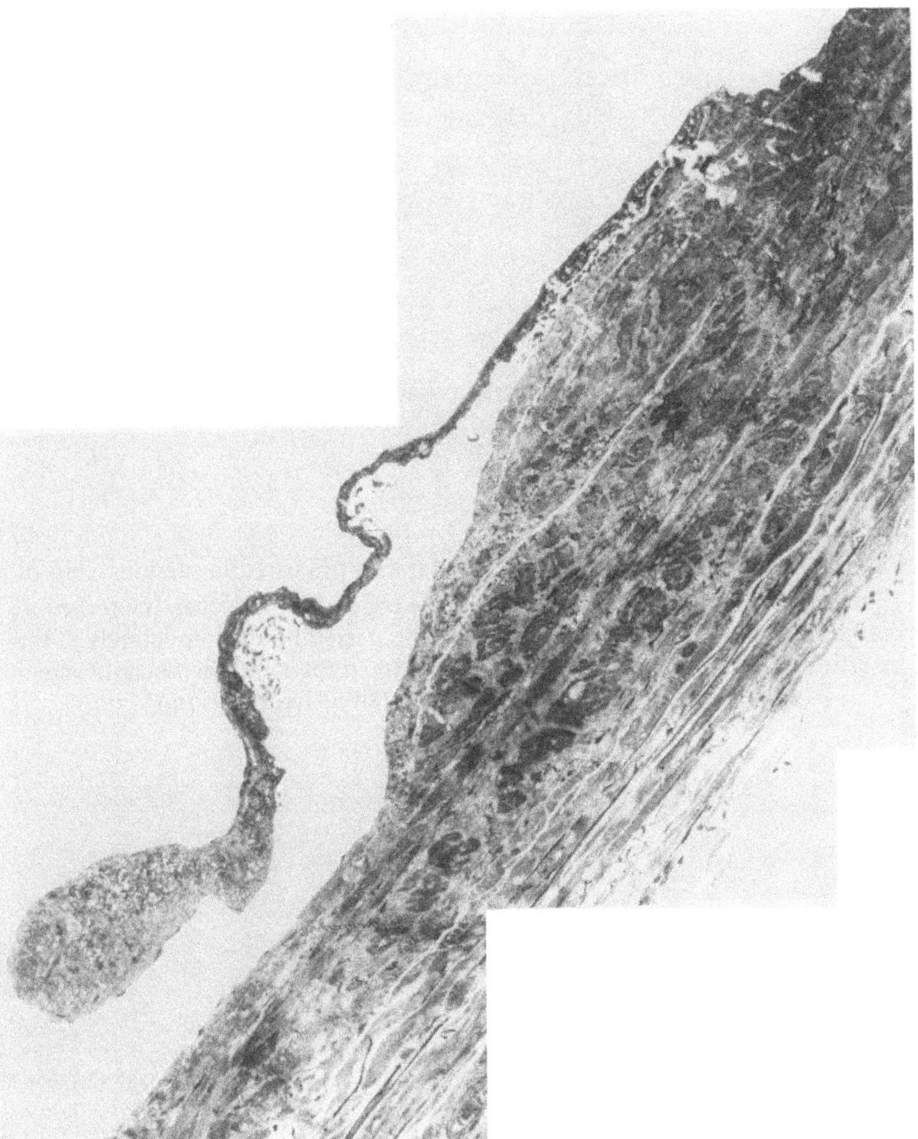

Fig. 18. Semi-thin section. Longitudinal section through valve of saphenous vein. 82-year old female. Tessellated picture. Note the beaded free edge of the valvular cusp

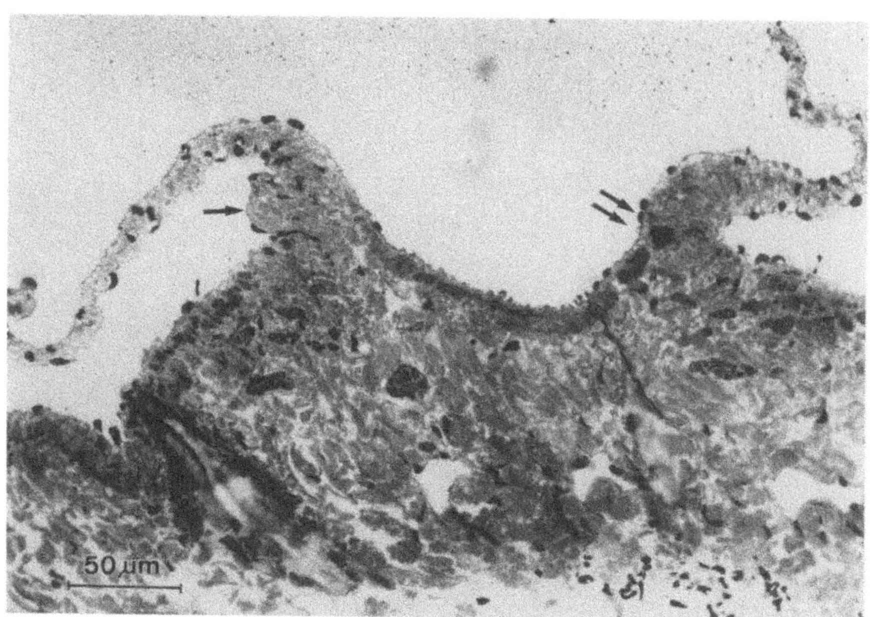

Fig. 19. Semi-thin section across valvular commissure. Saphenous vein of 18-month old infant. Note the beaded free edge of the left cusp (near the left margin of the picture) and the "fold" in the parietal part immediately at the insertion of the cusp (↑). The cusp on the right side shows only agger attaching the cusp to the venous wall at its insertion (↑↑)

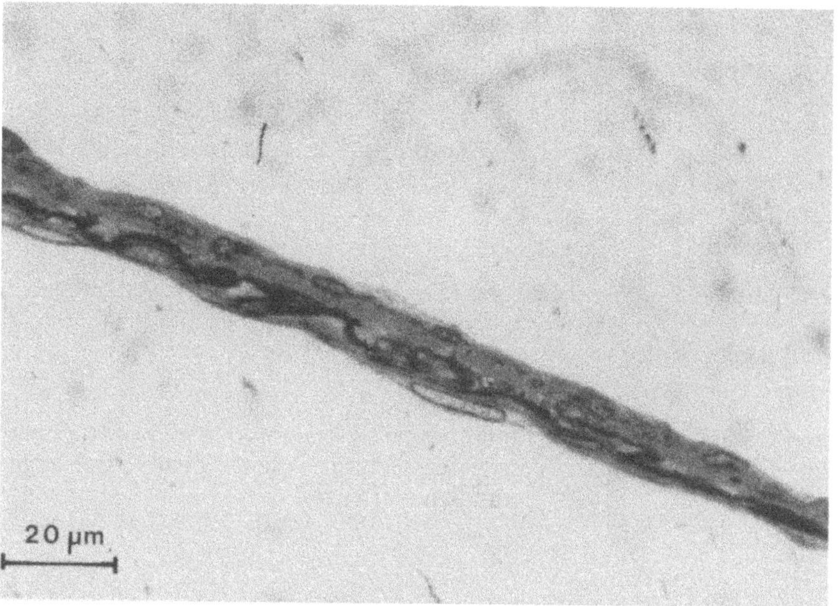

Fig. 20. Semi-thin section. Valvular cusp in internal saphenous vein of 34-year old male

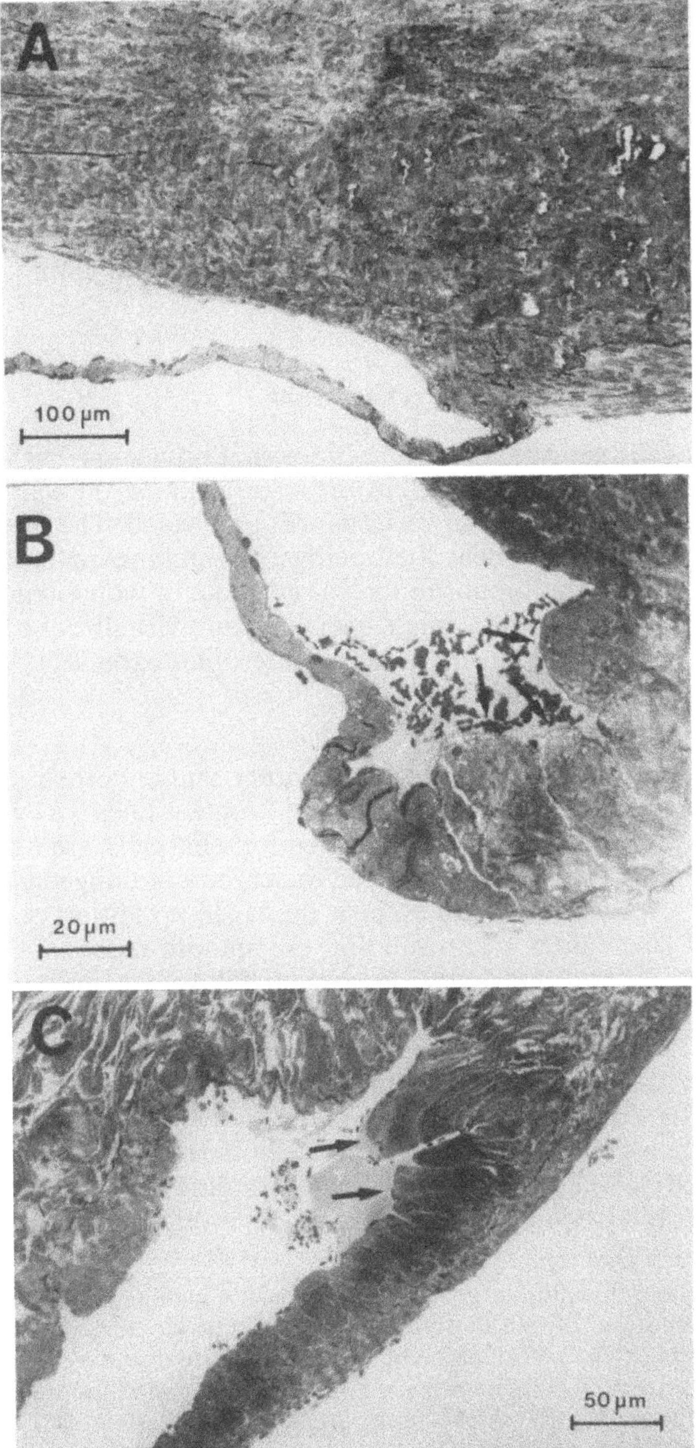

Fig. 21. Semi-thin sections. **A** Insertion of valvular cusp. Clearly visible agger, no folds in parietal part of cusp near insertion. **B** and **C** Marked folding of parietal part near insertion of valvular cusp (arrows). These folds may be identical with the buttress-like supporting structures visualized in en-face preparations (Fig. 24 B)

identical with it: they rather seem to be buttress-like folds occurring in large-sized valves for the purpose of contributing to the firm attachment of the cusp to the venous wall in the vicinity of the commissure. We presume that they are identical with the structures we found in en-face preparations (Fig. 24 B). Apparently it depends on the direction of the section as to whether or not such folds are hit upon.

c) Silver-stained En-face Preparations

It is one of the shortcomings of sections that only a narrow segment of the valve is visualized and, in order to examine the valve in its entirety, numerous series of sections will be needed. The method of "en-face preparation" described below, in combination with silver staining, offers the opportunity to evaluate a valve under the microscope by means of a single specimen and visualize the various layers of the valve by means of a few microphotograms.

1. Silver Staining

It has been known for more than a century that endothelia may be visualized by silver nitrate staining. The "cement lines", a system of silver lines marking the borders of the endothelial cells, are stained. If the staining silver solution is allowed to act for a prolonged period of time, or if the endothelium is in a damaged condition, a second system of silver lines – surrounding the smooth muscles – can be visualized. Each of these silver lines, which have become part of the media, will surround a smooth muscle cell and mark it off laterally from the adjacent muscle cell. Thus a system of predominantly horizontal networks of silver lines will develop in the mounted vessels. *Gottlob* and *Hoff* (1967) proved that the endothelial silver lines in the region of the fenestrae of the internal elastic membrane are related to the medial silver lines running along the smooth muscles. It is a frequent observation in whole-thickness preparations

Fig. 22. Whole-thickness preparations of rabbit's jugular vein. **A–B–C** Increasing duration of silver staining. **A** Endothelial cement-lines with spur-like processes. Occasionally the cement-lines are distorted at the origin of spurs (arrow). **B** When subjected to a longer period of staining the specimen shows that the spurs actually are incompletely stained "transverse lines" running along the smooth muscle cells of the media. **C** The transverse lines have been almost completely stained. The cytoplasm of the endothelial cells has been stained, the cell nuclei have been largely spared. The nuclei are surrounded by a thin, partly stained membrane

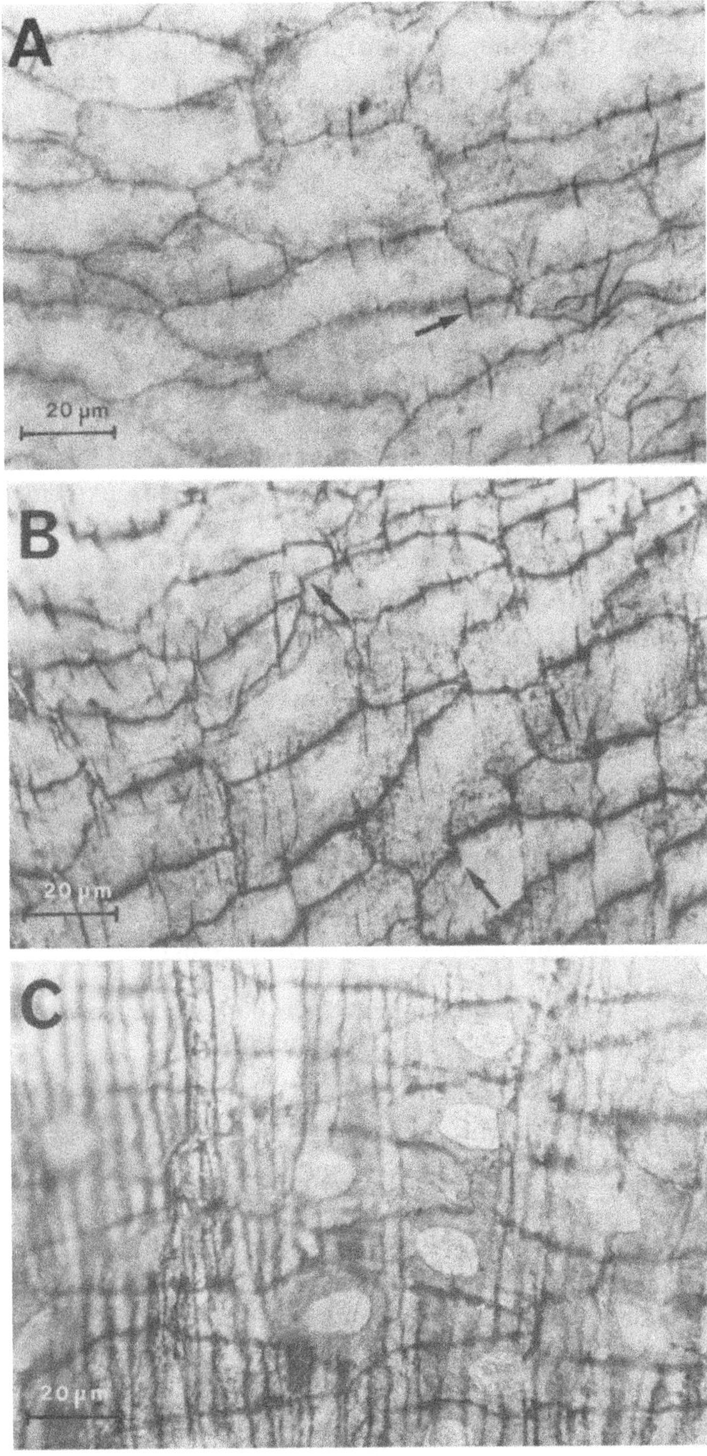

that the endothelial silver lines are distorted in the zones of adhesion with the medial silver lines. These findings led to the conclusion that there actually is some sort of "cement" anchoring the endothelium to the smooth muscles of the media in the region of the fenestrae (*Gottlob* and *Hoff,* 1967, Fig. 22, A, B, C).

For quite some time it remained utterly mysterious which organic substrate the silver lines corresponded to. Occasionally the lines are of granular appearance; their thickness is considerably greater than the intercellular space between the endothelial or smooth muscle cells. By histochemical studies it was proved (*Gottlob* and *Hoff,* 1968) that indeed we seem to be dealing with a cement-like substance, which is situated in the intercellular space and is so small that it will evade visualization by light-optical means. *Luft* (1966), however, found evidence of an intercellular substance by electron-optical methods (magnified 120.000 times). So why is this extremely thin layer of a cement-substance visualized by silver staining? The process involved is two-phased and not unlike that of photography.

During the *first phase* microscopically invisible silver grains form. During the first phase of photography, metallic silver is released from the silver bromide emulsion due to the action of light. The silver particles released in the photographic process are also microscopically invisible. During a *second phase* there is massive accumulation, by aggressive reducing action of the developer, of silver to the silver grains pre-formed during the first phase – an image visible by light-optical means evolves. – In endothelial silver staining the effect of the silver nitrate upon the endothelial cells results in the formation of silver halides in the tissue, which primarily contains chlorides. In no other region than in that of the minute endothelial cement substance, is metallic silver reduced and bonded primarily. The second phase of endothelial silver staining is that of the action of daylight, by which metallic silver is reduced from the silver halides in the tissue. The metallic silver then deposits on the pre-formed and hitherto invisible particles so that a system of cement-lines visible by light-optical means appears.

2. En-face Preparations

To avoid the drawbacks inherent in sections due to the small quantities of tissue available for examination, *O'Neill* (1947) for the first time studied silver stained vessels in the form of whole-thickness preparations (en-face preparations). For that purpose *O'Neill* cut open the vessels, spread them with the intimal layer facing upward and, after silver staining and dehydration in alcohol solutions of increasing concentration, cleared them up in wintergreen oil. The vessels were mounted between slide and coverglass and offered themselves to evaluation of endothelial areas in their entirety. – This method of "en-face preparation" has proved its worth in particular for studies of endothelial pathology. *Zinner* and *Gottlob,* e.g., used the method for proving damages caused to endothelium by contrast media (1959). – The method of endothelial silver staining and full-thickness en–face preparations was first used for investigating venous valves by *Gottlob* and *Kimmel* (1973).

3. Method

The veins whose valves are to be studied must be dissected with all due care; any contact of the vessel with forceps, hand or swabs should be avoided. Only the surrounding connective tissue should be seized, and severed with sharp scissors.

Silver staining may be applied in situ – either at the intact or at the opened vessel. For silver staining *in situ* the vessel is dissected or ligated proximally and distally to the valve, cannulized by a fine plastic catheter proximally to the valve, cut open distally to the valve and flushed with saline or buffered Ringer solution until all blood has been drained from it. The vessel is then flushed briefly with a 5% solution of glucose in water, and after that, for 20 seconds with a 0.25% silver nitrate in water solution. This is followed by another brief flushing with a 5% solution of glucose in water, after which the vessel is subjected to fixation by rinsing with a 5 to 10% formalin solution, which may be achieved by application of a continuous-drip infusion for several hours. During silver staining, in order to make sure that both surfaces of the venous valves can be safely visualized, the solution may be re-aspirated briefly so that there will be a temporary retrograde flow.

After staining and fixation, the vessels are dissected in their entirety, spread on cork sheets approximately in their natural length, and cut open using a fine pair of scissors, proceeding in the direction of the bloodflow. The vessels are then spread out by means of fine needles, avoiding any damage to the areas of valves when sticking the needles in. Thereupon, the vessels are dehydrated in alcohol solutions of increasing concentration, immersed in wintergreen oil (methylsalicylate) for one hour and cleared. After that, the specimens may be mounted between slide and coverglass in the form of whole-thickness preparations in Caedex or some other balsam. This is the safest method of preserving the intactness of the endothelium.

As an *alternative,* the vessels may be dissected, cut open and spread prior to staining, in which case the exposed endothelium is stained by gently sprinkling it with silver nitrate solution. Before and after that it is flushed in a 5% solution of glucose in water. Care should be taken to apply the solutions in the opposite direction of the bloodflow in order to make sure that they will well penetrate the valvular sinus. In the same manner as with the intact vessel, the specimens are dehydrated, and mounted.

4. Results

As a rule 3 layers of endothelium can be observed in the regions of *valvular cusps:* the endothelium of the venous wall and two layers of valvular endothelia (Figs. 23, 25, 26).

In the avulvar portions of the vein the endothelium of the *venous wall* predominantly runs in such a way that the longitudinal axes of the cells are aligned in correspondence with the axis of the vessel. The arrangement in the region of the *valvular sinus* is different. There the endothelia is arranged either crosswise to or in irregular alignment with the axis of the vessel. The same applies to the endothelium of the parietal part of the valvular cusp. In contrast thereto, the luminal part of the cusp shows an alignment of its endothelia, which is once again predominantly in correspondence with the axis of the vessels. – It has been known that endothelial cells tend to align their longitudinal axes with the direction of the bloodflow. This principle is fully upheld in the region of the valve. When the valve is open, only the luminal part is subjected to the bloodstream – which is why the endothelial cells are aligned in longitudinal direction. In the valvular sinus, however, there is no considerable bloodflow, regardless of whether the valve is open or closed. The endothelial cells, therefore, are arranged in an irregular manner and frequently crosswise to the axis of the vessel. On the other hand, we would argue against any temptation to assume that the crosswise positioning of the endothelial cells were "lawful" in any way. Especially in en-face preparations one may realize that their arrangement is of considerable irregularity and thus in correspondence with the turbulent flow conditions in that region.

In a comparison of the endothelial cells of the venous wall with those of the valvular cusp it is striking that frequently the silver lines of the venous wall are stained much more intensively than those of the valvular cusp. – Apart from that there are no other microscopical-

Fig. 23. Venous valves in en-face preparations. Silver staining. Region of the edge of the valvular cusp. Jugular veins of a rabbit (left) and a dog (right). **B** and **D** Continuous endothelial layer of venous wall, **A** and **C** Layers of valvular cusps with two systems of cement-lines; in each of them only one system of cement-lines runs horizontally (parallel with the axis of the vessel), this is in both of the cases, the endothelium of the luminal part

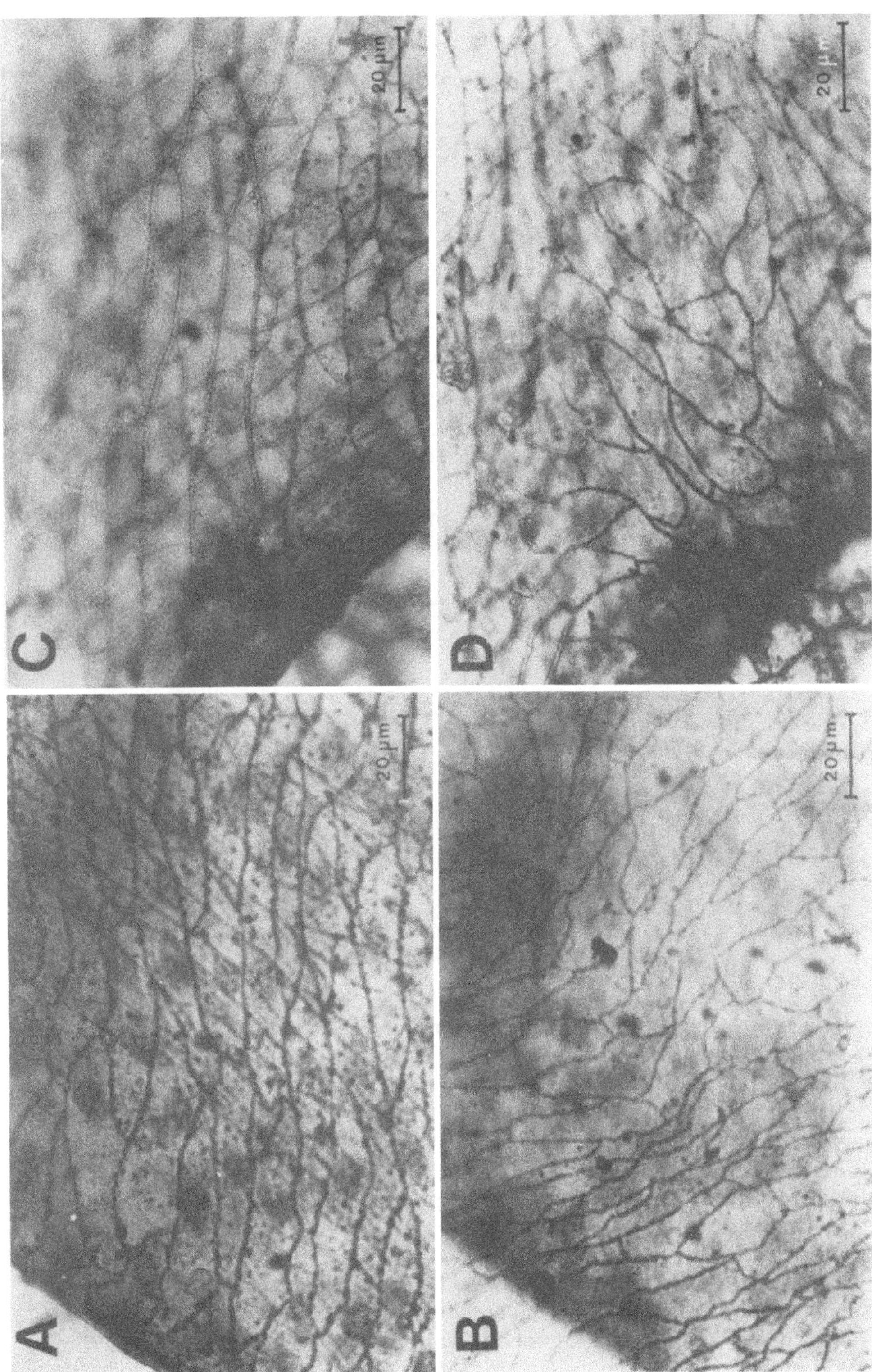

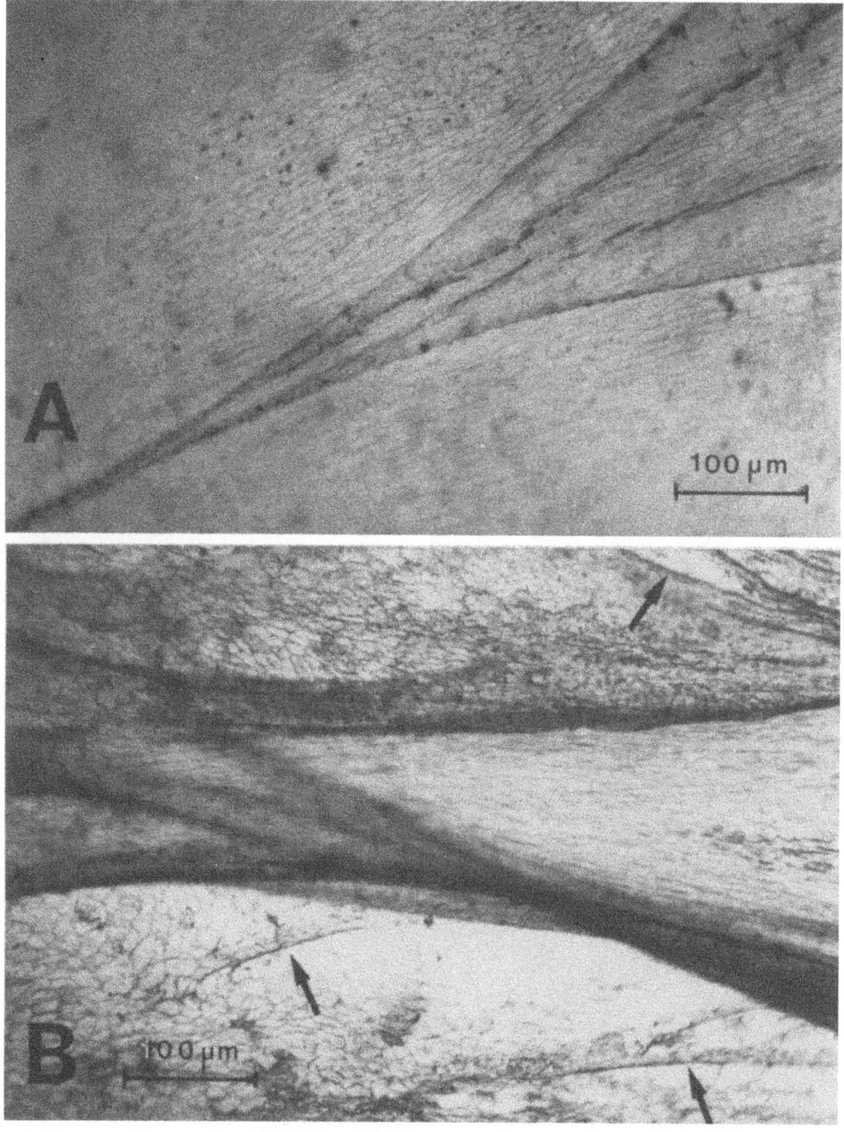

Fig. 24. En-face preparations, silver staining. Commissures of venous valves. **A** Rabbit. The undamaged commissure is acute-angled. **B** Human saphenous vein. In large veins the valves are less acute-angled; in addition thereto, buttress-like folds (arrows) extend from the commissure into the parietal endothelium of the vessel

Fig. 25. Distal end of a valvular sinus. Jugular vein of dog. Same area shown at three different levels. **A** Uppermost layer (near lumen). In the lower right corner, endothelium of venous wall, merging into the endothelium of the luminal part of the valve (more to the left). **B** Left, endothelium of the parietal part of the valvular cusp. In the right lower corner the endothelium of the vessel distal to the valve is visible. **C** Left side and upper left corner, endothelium of venous wall in the region of the valvular sinus

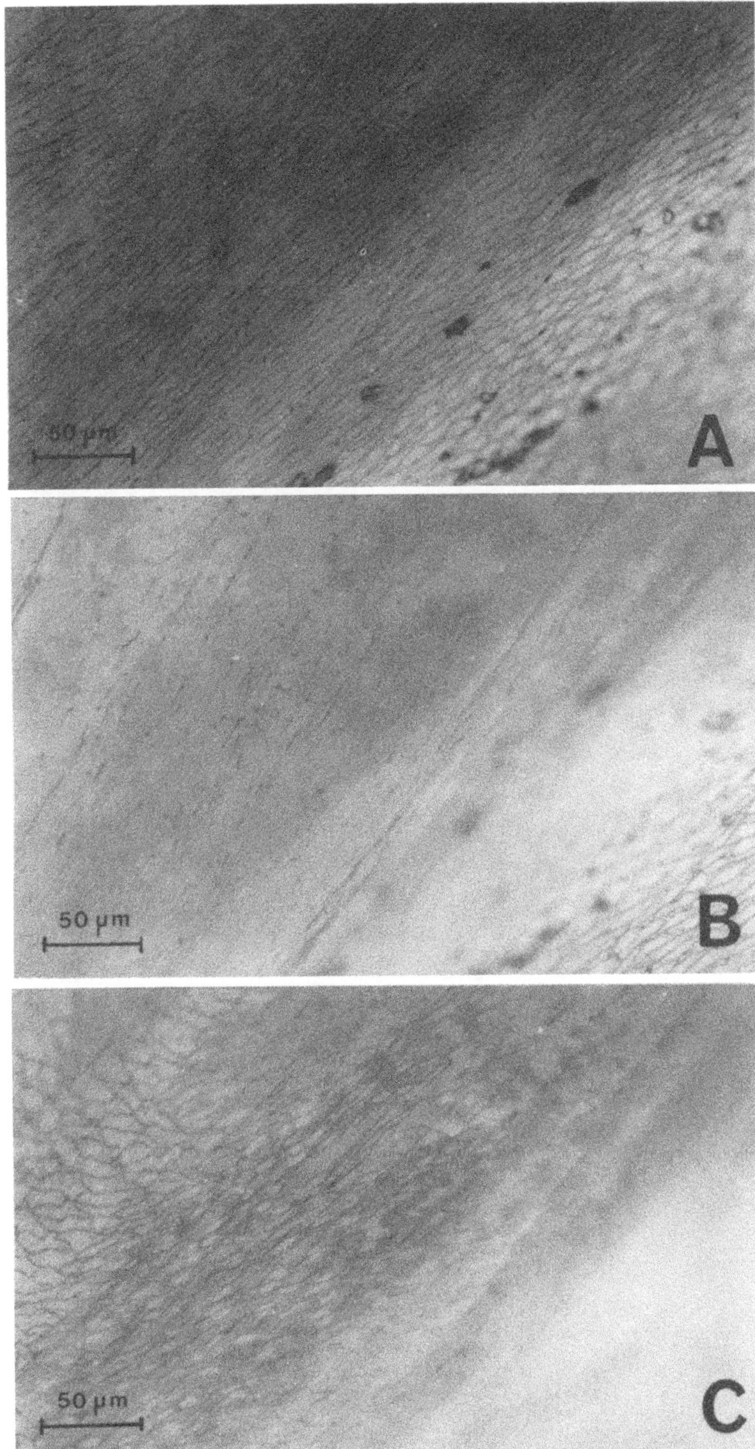
50 µm
A
50 µm
B
50 µm
C

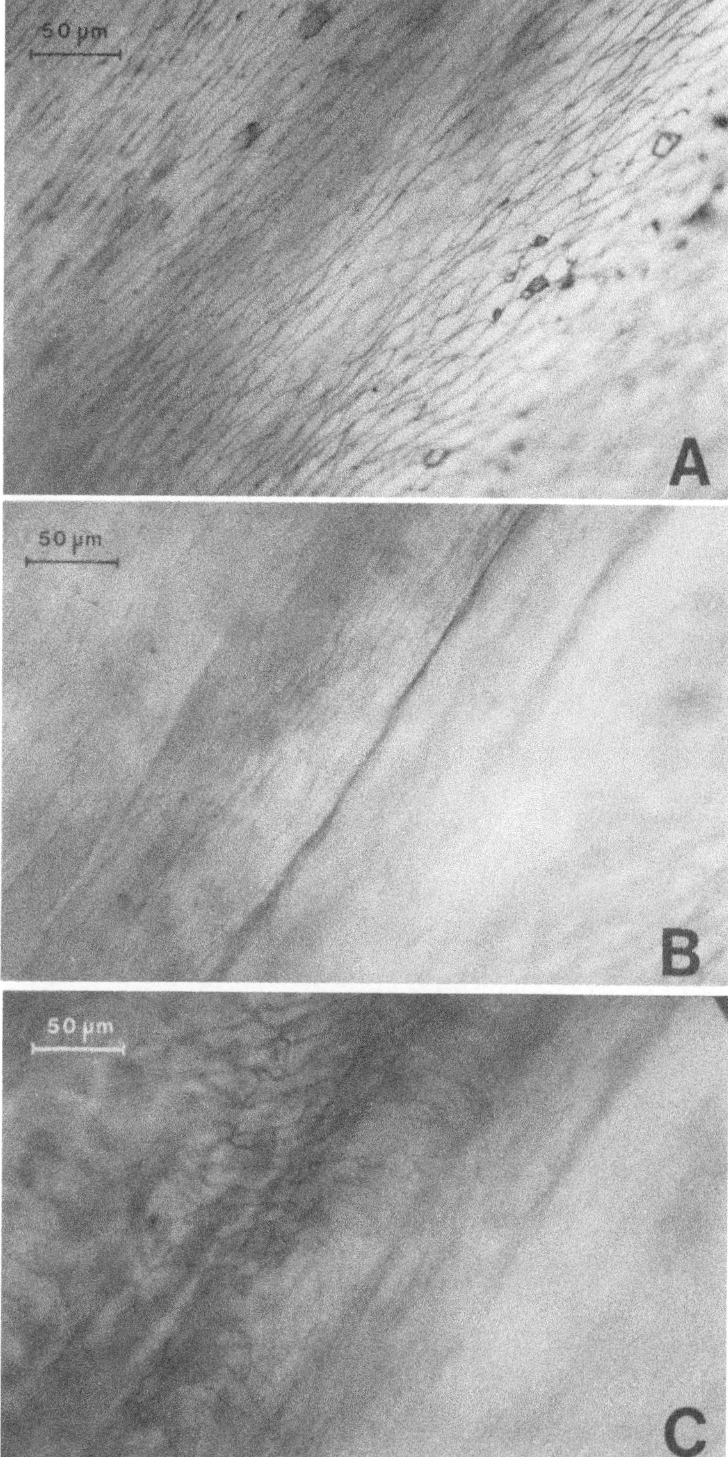

ly conspicuous differences between valvular endothelia and endothelia of the vessel wall in specimens taken from animals (rats,
rabbits, dogs). The same applies to specimens taken from humans of
lower age groups. In older human individuals considerable differences are found. We shall revert to those differences when discussing
changes caused by age (pp. 88–92).

Fig. 24 shows the *commissures* of venous valves taken from rabbit
and human material. A normal and undamaged valvular commissure
is acute-angled with smaller valves and somewhat wider with larger
valves; in the region of the commissure the larger valve is anchored to
the parietal endothelium by several folds of buttress-like appearance.
We assume that these folds may be identical to the ones we have
observed in semi-thin sections (Fig. 21 B and C).

Figs. 25 and 26 show the *edge* of the *valvular sinus* located at the
"windward" side of the bloodflow. – On the free edge of the valvular
cusp located at the "leeward" side of the bloodflow we find an
endothelial layer in continuous course lying at the bottom, i.e. on the
wall of the vessel. At the distal end of the blind sac, located at the
"windward" side of the bloodstream (Figs. 25 A and 26 A), we find
continuous endothelial lining in the uppermost layer of the specimen,
i.e. on that side of the valve facing the lumen.

Figs. 27 and 28 demonstrate the excellent quality by which ostial
valves may be visualized in en-face preparations.

For the purpose of evaluating whether a venous valve is damaged
or not, *three layers* of cement-lines should be visible in the valvular
region. Often the endothelial layer of the parietal side of the valve is
stained only slightly. In view of that fact, we may say from experience
that, as a rule, two endothelial layers clearly located on top of each
other will be sufficient to confirm the inference that the valve in the
region concerned is not adherent but equipped with a free cusp.
Images of pathologic venous valves will be discussed in the section
dealing with pathologic histology.

En-face preparations of silver stained venous valves offer the
advantage of evaluation of the venous valve in its entire layout. The
method has its *drawbacks* too: first of all, the structures to be
investigated are arranged at varying levels, the visualization of which

---

Fig. 26. Valve in human internal saphenous vein. En-face preparation. **A, B**
and **C** Same area shown at three different levels. The distal edge of the
valvular sinus is visualized. **A** Uppermost layer facing the lumen; in the right
lower corner, endothelium of venous wall, merging (more to the left) into
luminal endothelium of valvular cusp. **B** Parietal endothelium of cusp (left).
C endothelium of venous wall in the region of the valvular pocket

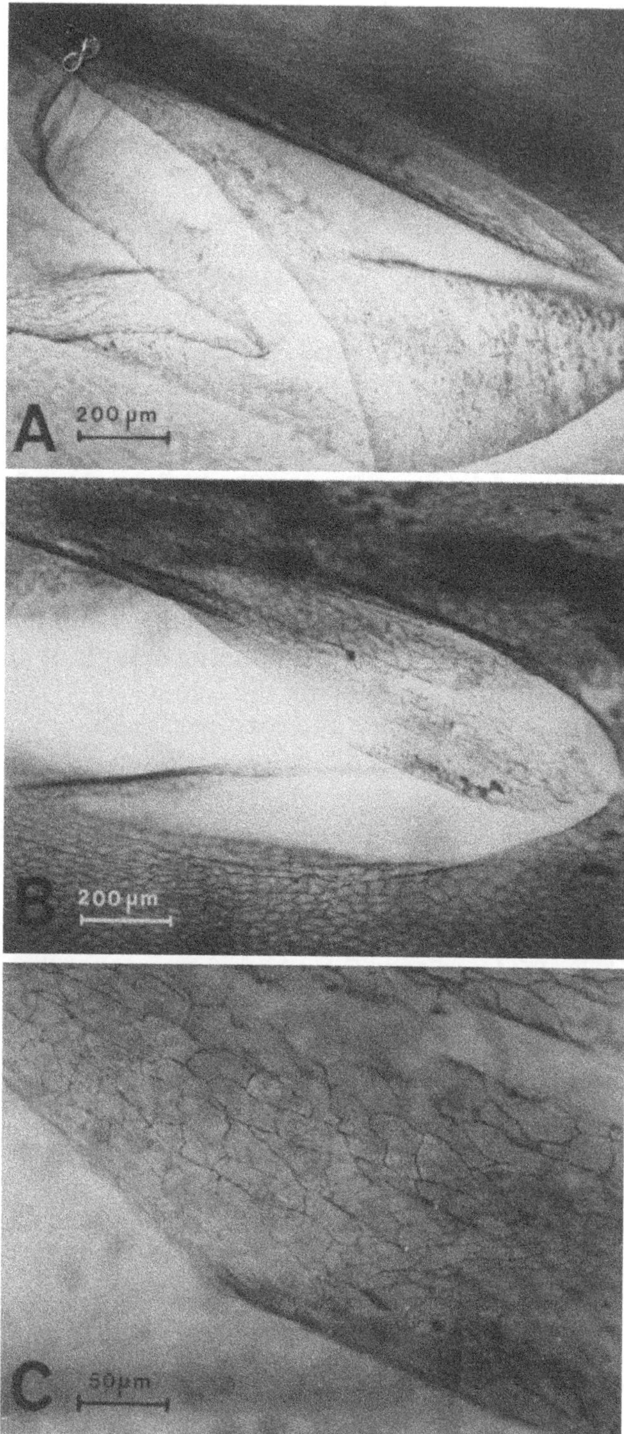

200 µm
A
200 µm
B
50 µm
C

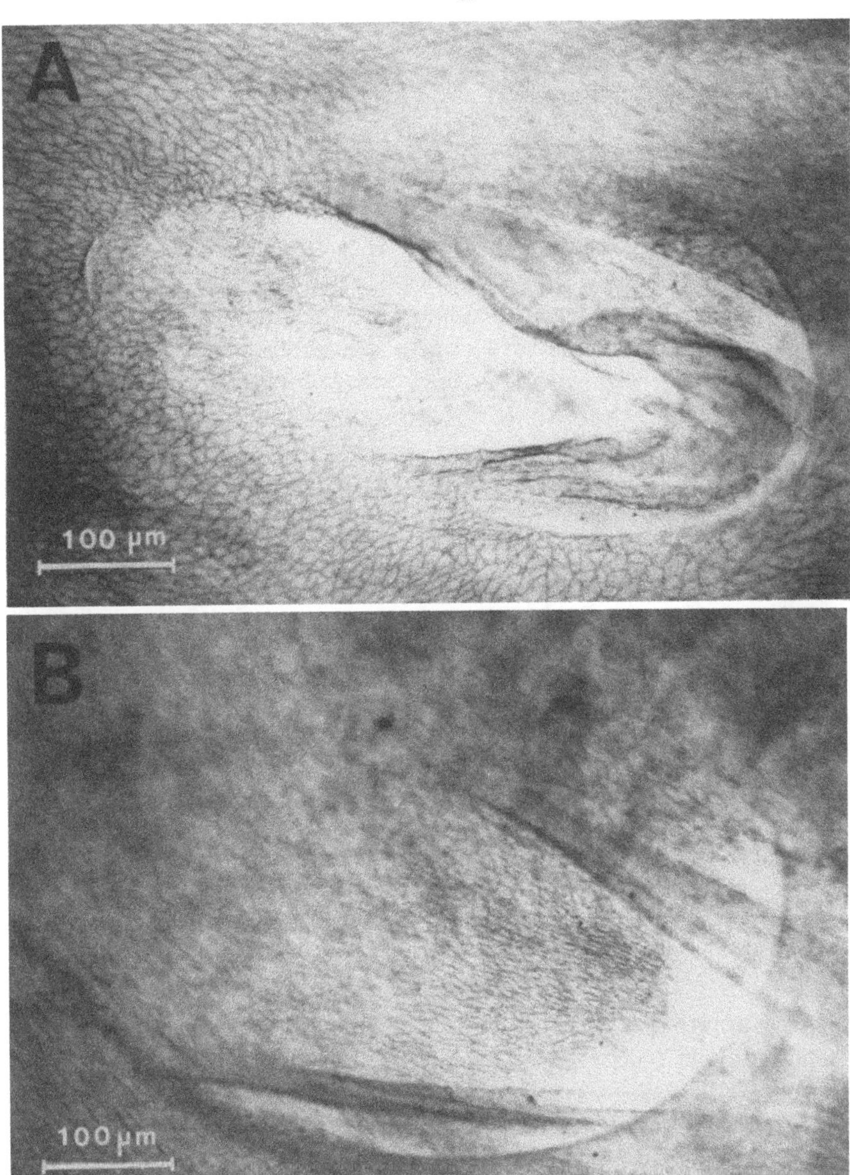

Fig. 28. Ostial valves with "recessed" insertion. **A** Unicuspid valve. **B** Bicuspid valve. Jugular vein of dog

---

Fig. 27. **A** Bicuspid ostial valve, "marginal" insertion. The cusps hang out freely into the lumen of the main vessel. **B** Unicuspid ostial valve. **C** Same cusp as shown in B, under higher magnification. Two endothelial layers are visible

is often far from optimal. Because of this, the person evaluating the
specimens is frequently required to readjust the micrometer screw of
the microscope. The disadvantage caused by multilayer arrangement
and uneveness of elevation is not likely to facilitate the use of
microphotographic methods either. Another drawback is that the
vessel must be cut open so that, as a rule, at least one cusp will be
damaged. – However, in view of the fact that some of these
drawbacks are inherent to conventional sections as well, and keeping
in mind that en-face preparations enable the investigator to evaluate
a venous valve by means of a single specimen, we do regard silver
stained en-face preparations a very suitable method of examining
venous valves as to their anatomic intactness. Besides, after examination of the endothelial lining and production of microphotograms,
whole-thickness preparations may easily be further processed to
conventional sections or monolayered flat-preparations.

d) Unicellular Flat-Preparations ("Häutchenpräparate")

Occasionally, in order to stain the nuclei of endothelial cells and
obtain histologic specimens as flat as possible, monolayered flat-preparations ("Häutchenpräparate") were produced according to the
method developed by *Lehr, Nashef* and *Gottlob*. For that purpose we
proceeded by first dehydrating the silver stained whole-thickness
preparations in alcohol solutions of increasing concentrations; then
one surface of slides made of methylmetacrylate ("Plexiglas") was
macerated in acetone for 30 minutes. The vessel was placed on the
slide, with the endothelial layer facing the macerated surface of the
slide; both vessel and slide were pressed between cork sheets in a vise
for one hour. Thereupon the vessel was drawn off the slide (like a
decal) in such a manner that the endothelium – in most cases – was
left on the slide in the form of a single-layered film ("Häutchen").
The specimens were stained by hematoxylline overnight, dehydrated
once more, and protected by a coverglass after application of a few
drops of Caedex.

The results obtained by the unicellular flat-preparation method
will be discussed in the chapter dealing with age-processes
(pp. 88–92).

e) Electron Microscopy

By S. *Geleff* *

1. Method

For electron microscopy the sections were prepared in the same manner as for semi-thin sections. The samples embedded in epon 812 were processed to sections whose colors ranged from gold to silver, and collected on uncoated copper grids. Contrasting was achieved by a 25% uranylacetate solution for eight minutes and lead-citrate for two minutes. Tannin-contrasting was achieved according to the method developed by *Kajikawa et al.* (1975) as modified by *Stadler* and *Orfanos.*

The sections were examined under a Zeiss EM 9 electron microscope.

2. Results

2.1. Cellular Elements

2.1.1. Endothelium

The parietal and luminal parts of the free portions of valves are covered by uni-layered endothelial cells (cf. Fig. 29 A).

In the endothelial cells of both human and animal venous valves all organelles and cellular elements typical of such cells are found. In particular there are a large number of transport vesicles and "coated vesicles". The number of Weibel-Palade bodies varies (Fig. 29 E).

As in the venous wall itself, the endothelial cells are attached to each other by "closed junctions". The superficial borders of cytoplasm of both the parietal and luminal layers of endothelial cells of the venous valve are gathered in overlapping marginal folds in a regular manner:

In almost all of the cells large quantities of filaments, the diameters of which corresponde to those of the "intermediate filaments", are to be found (cf. Fig. 29 B). Occasionally there is also evidence of a single cilium.

The distance between the endothelial cells of the luminal and parietal sides varies. Frequently those cells are so close to each other that they are separated by only a very few collagenous fibrils or none at all. The basal membranes of such processes of cytoplasm may

---

* These studies were made at the Institute of Micromorphology and Electron Microscopy of the University of Vienna. All authors feel greatly obliged to Prof. Dr. P. *Böck* for his advice and valuable suggestions.

touch or even merge with one another (cf. Figs. 29 A and 30 B). In contrast, the distance between the two layers of endothelial cells may be quite considerable at times (cf. Fig. 29 A), and either one or more bundles of collagenous fibrils, or one or more connective-tissue cells (myofibroblasts) may be interspersed between them (cf. Fig. 29 A).

2.1.2. Stroma Cells

In the connective tissue of the free valvular portion stroma cells are present in sparse quantities. They may appear approximated to one or the other side of the endothelium or, by their processes, reach up to the endothelia of any of the two valvular surfaces (cf. Fig. 29 A). Numerically, they are present in the valvular cusp in varying quantities. In the free edge of the valvular cusp they occur in larger quantities, are markedly ramified and cohere to each other by means of their processes (cf. Fig. 32). Two types of stroma cells are

---

Fig. 29 A – E.. Valve from jugular vein of rabbit
A Low-power magnification. The valvular sinus on the left side still contains erythrocytes and fixed plasm; the side facing the lumen of the vessel has been flushed by fixing solution. The valvular stroma, consisting of collagenous fibrils *(k)*, is of varying thickness. In circumscribed areas, the endothelial cells of both surfaces touch each other *(b,* cf. also Fig. **C).** In the figure, two stroma cells are visible *(S₁, S₂).* The valvular surface is coated by regularly built endothelial cells (magnified 4,200 times)
B Detail of a stroma cell (S₂ in Fig. A). The polygonal cells, which frequently develop far-reaching and ramified processes, appear to be fibroblasts. In some cases bundles of cytoplasmic filaments *(f)* and fragments of a basal lamina *(bl),* as have been described for the contractile form of fibroblasts (myofibroblasts), are found. The short profiles of the rough endoplasmatic reticulum are expanded and filled with a modestly electron-dense granular substance. The diameters of the collagenous fibrils in the valvular stroma range between 30 and 50 nm (magnified 25,000 times)
C In valvular regions devoid of stroma, the endothelia of boths sides are closely adjacent to each other. The basal laminae merge (arrow) (magnified 28,500 times)
D Sparse elastic material is found predominantly at subendothelial sites. By tannin and uranylacetate contrasting it may be visualized as electron-dense (arrow) (magnified 4,000 times)
E The endothelial cells are characterized by transport vesicles *(t)* and Weibel-Palade bodies *(wp).* They rest on a basal lamina, to which bundles of "elastic fiber microfibrils" are (frequently) aggregated (arrow, cf. also Fig. 30 A). In the cytoplasm of the endothelial cells, numerous "intermediate filaments" are visible. Most of the cells border on each other by "closed junctions", with their superficial edges gathered and forming overlapping marginal folds (magnified 30,000 times)

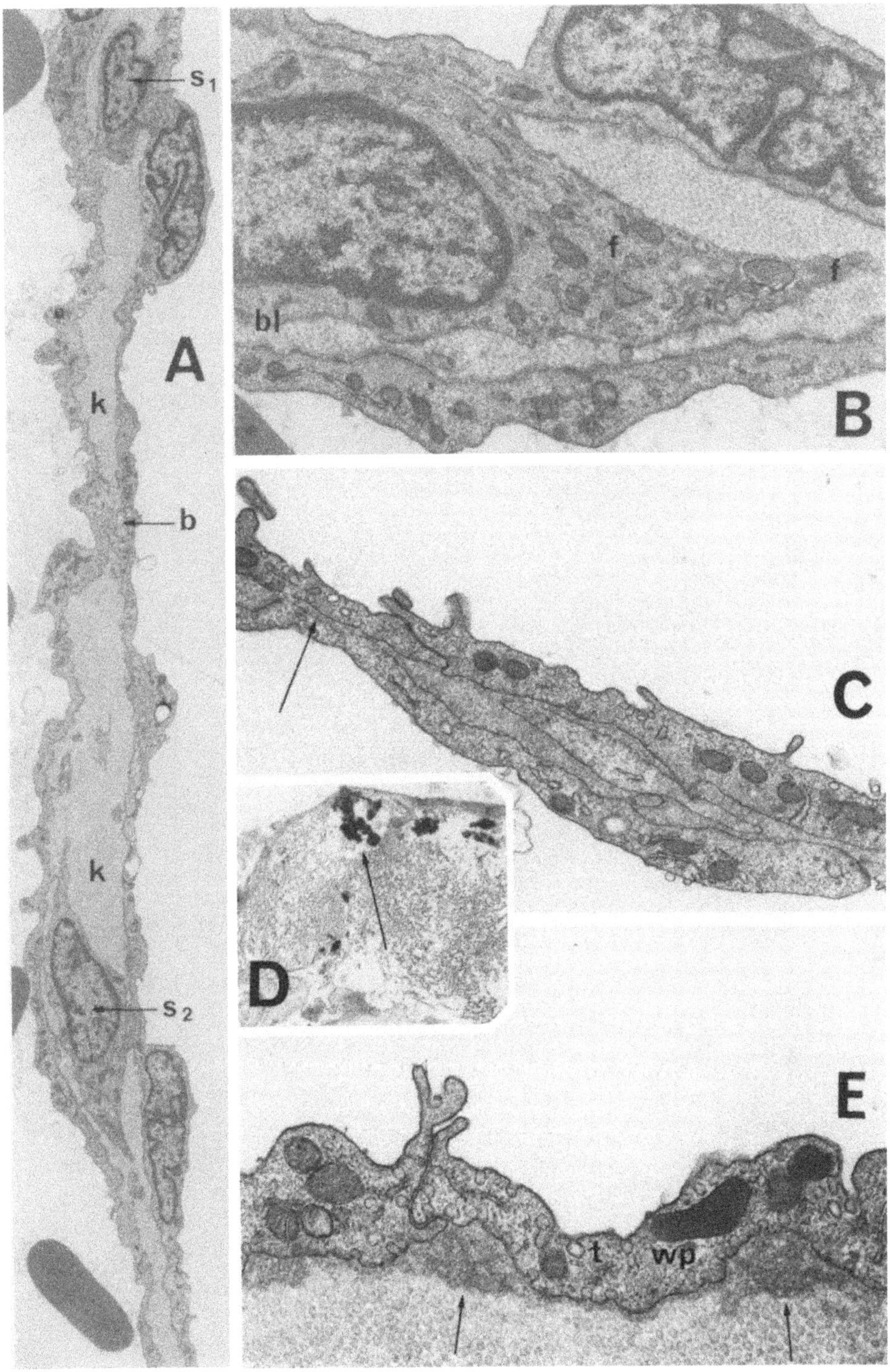

s₁
k
A
b
k
s₂
bl
f
f
B
C
D
E
t
wp

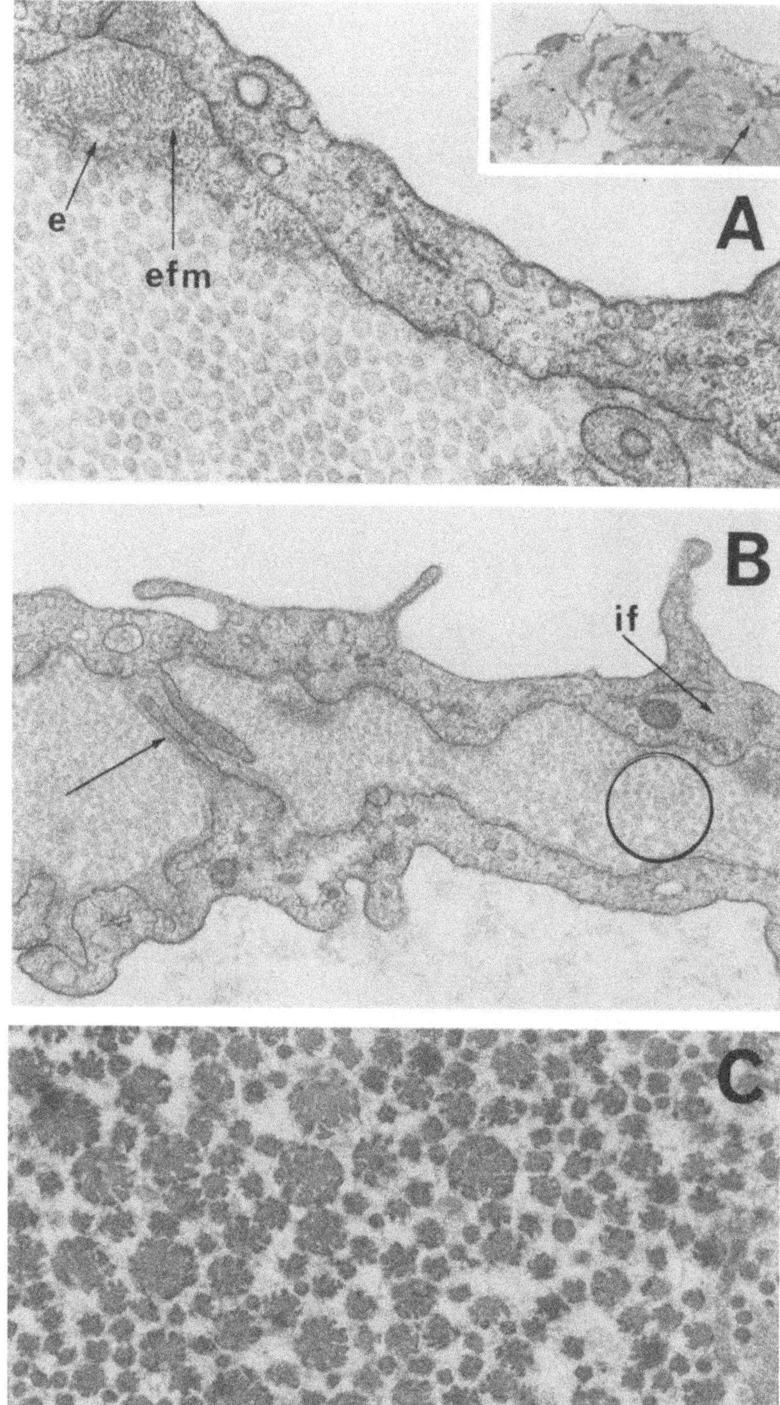
e
efm
A
B
if
C

observed, which are morphologically distinguished from the fibro-blasts and fibrocytes usually found:

a) Connective-tissue cells, located slightly below the endothelium and showing a discontinuous basal membrane. In addition to that, these cells show bundles of cytoplasmic filaments (the diameter of which is approximately 80 Å) and, at their surface, pinocytotic vesicles.

b) The second type of cells, which also occurs predominantly at subendothelial sites, is similar to endothelial cells and, like those, is provided not only with bundles of equally sized filaments but also with a number of transport vesicles, and a partially developed basal membrane. Frequently these subendothelial cells give the appear-ance of "light" cells (cf. Fig. 31 B).

c) Smooth muscle cells with characteristic bundles of myofila-ment. "Dense zones" and basal membranes are extremely rare findings. For that reason it is difficult to judge how far, from the insertion of the cusp (where they may be found in single small groups), they reach into the free portion of the cusp.

2.2. Fibrillar Elements

2.2.1. Collagenous Fibrils

The main portion of the fibrillar material in the venous valve is represented by collagenous fibrils.

In "thin" areas of the valve, usually only one layer of fibrils, running in a parallel course and combined to a bundle or a lamella, is

---

Fig. 30 A–C

A Under light-optical microscopy (insert) positively stained elastic fibers (arrow) can be detected in sparse quantities by classic methods of elastica staining (in this case by Weigert's method of resorcin-fuchsin staining modified by Romeis). Under the electron microscope it is obvious that these fibers consist of small quantities of elastica (e), predominantly representing bundles of "elastic fiber microfibrils" (efm) (magnified 45,000 times)

B The endothelial cells of both surfaces of the valve may develop cytoplasmic processes, reaching into the depth of the stroma and meeting each other (arrow). A bundle of "intermediate filaments" corresponding to those shown in Fig. A has been marked (if). Not only do the collagenous fibrils of the stroma frequently vary in dameter – but thicker fibrils may even show irregularly outlined cross-sections (area marked by circle, cf. Fig. C) (magnified 28,500 times)

C Collagen from wall of human saphenous vein; age 70. Under electron microscopy the macroscopically inconspicuous vessel shows the marked picture of so-called "fibrillar dysplasia" of the collagen; there were entirely similar findings in the valvular stroma (magnified 28,500 times)

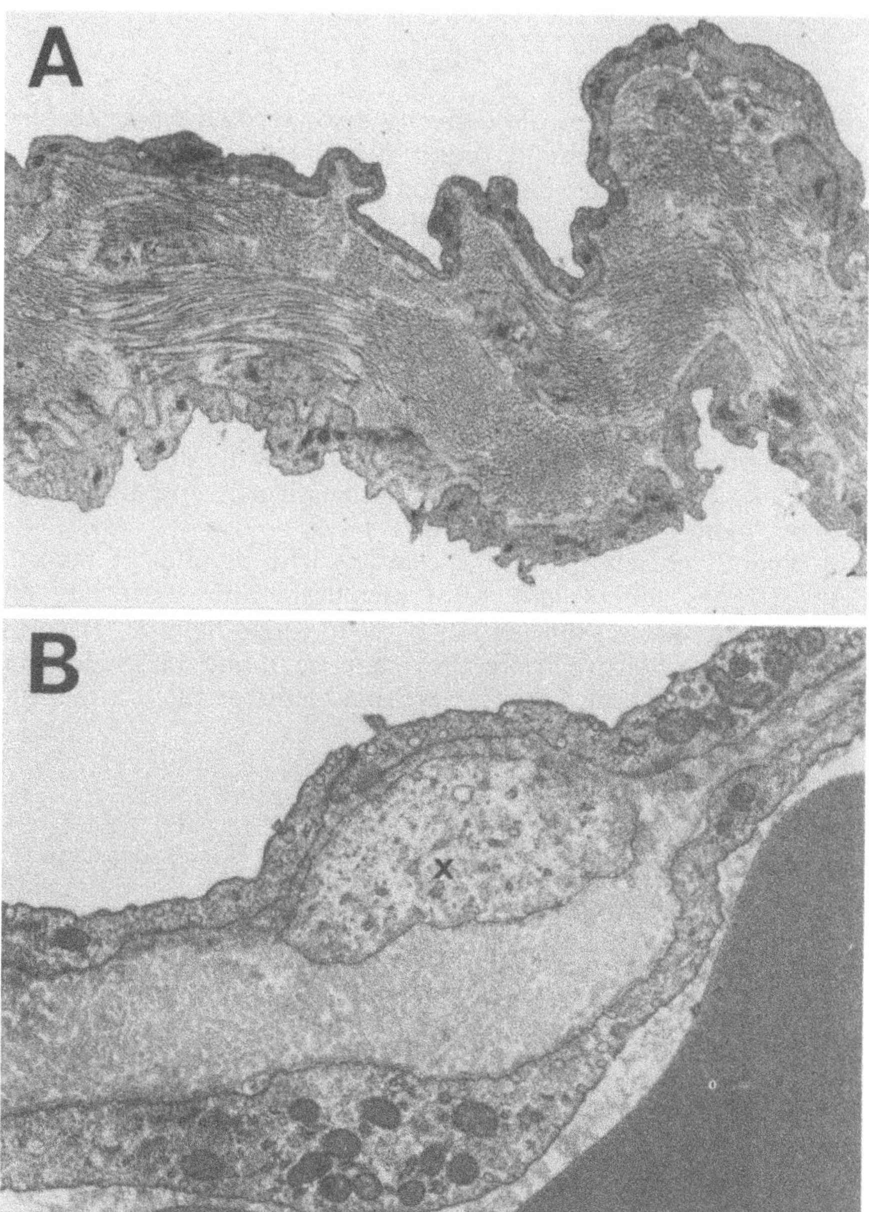

Fig. 31 A and B

A In thicker regions of the stroma the collagenous fibrils are combined to bundles and lamellae. Within these bundles the fibrils run strictly parallel; the alignment of the bundles toward each other cannot be discerned in single sections (magnified 7,000 times)

B Immediately underneath the endothelium, in particular under the cellular borders, cells with conspicuously light-colored cytoplasm *(x)* are found; they distinguish themselves from sections of stroma cells (myofibroblasts) by their number of organellae and cytoplasmic filaments. Such profiles were interpreted as "replacement cells" for the endothelium by Hoff and Gottlob. In the lower right-hand side of the picture an erythrocyte is adjacent to the endothelium (magnified 25,000 times)

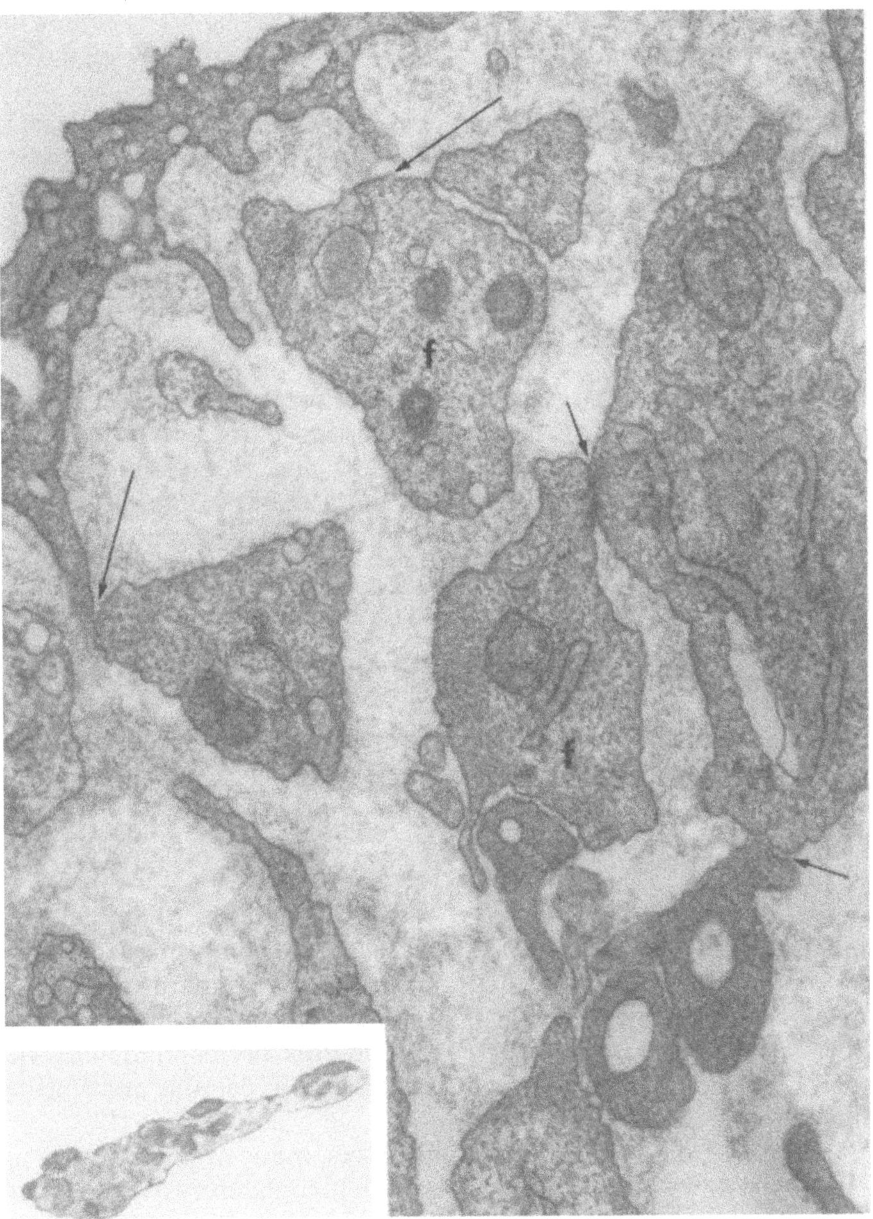

Fig. 32. Venous valve taken from rabbit. Detail close to free valvular edge. In the stroma of the thickened free edge of the valve (cf. insert – light-optical microscopy) ramified myofibroblasts (stroma cells) are found in increased quantities. The processes of these cells contain portions of the rough endoplasmatic reticle and bundles of cytoplasmic filaments (f). The processes touch each other (short arrows) as well as basal processes of the endothelial cells (long arrows) without developing specific zones of contact (magnified 28,500). (Light-optical microscopy in insert – magnified 700 times)

found. If the space between the two endothelial layers is somewhat wider, two or more bundles of collagenous fibrils crossing each other are to be found (cf. Fig. 29 A).

Within one bundle of fibrils there are variations of the diameters of fibrils (ranging between 350 and 600 Å). Upon closer examination collagenous fibrils with irregularly orbed, or even serrated, cross-sections can be observed. Such fibrils (cf. Fig. 30 C) have been referred to by the term dysplastic fibrils.

2.2.2. Other Fibrillar Elements

Using methods of both conventional and electron microscopy very small quantities of fibrils reactive to elastica-stains may be detected in the free portion of the valve. Close to the basal membrane of endothelial cells, as well as in the stroma of the valve, bundles of "elastic fiber microfibrils" are present (cf. Figs. 29 E and 30 A).

Tracing the internal elastic lamina of the venous wall, one realizes that there is no connection to the sparse elastic fibers in the luminal part of the valve.

After *tannin* contrasting, small and irregularly outlined electron-dense areas of elastin can occasionally be visualized, mostly at the luminal and parietal sides of the stroma and, less frequently, amid the bundles of collagenous fibrils (cf. Fig. 29 D).

3. Discussion

So far there have been very few reports on studies of human and animal venous valves by methods of electron microscopy; the only paper presented in more recent times is *Böck's* comparative study of arterial and venous valves (1975). A comparison of human valves with those of the rabbit or the dog reveals correspondence with respect to both their structure and the layout of cellular and fibrillar elements.

As to the connective-tissue cells of valves, many of them – by virtue of bundles of cytoplasmic filaments, which might correspond to actin filaments – may be considered myofibroblasts.

The second type of cells differing from fibrocytes usually found, is that of subendothelial "light" cells. Such profiles were also described by *Hoff* and *Gottlob* (1968) and, by methods of light-optical micros-copy, interpreted as indications of endothelial regeneration. The quantities of smooth muscle cells with continuously developed basal membrane, myofilaments and characteristic "attachment zones" are negligible. In view of the fact that per section no more than one smooth muscle cell, if any, will be found, the functional significance of these contractile elements is surely small.

The free edge of the valve, which frequently appears beaded even under light-optical microscopy, shows more connective-tissue cells than does the valvular leaflet under electron microscopy. Merely in the region of the valvular insertion do myofibroblasts and smooth muscle cells occur with comparable frequency.

The varying diameters of collagenous fibrils, which were observed in almost all of the valves examined, and the frequently found cross-sections of fibrils that appeared extremely thickened and serrated corresponde to those collagenous fibrils termed dysplastic by *Fischer* and *Staubesand* (1982). We would point out the fact, however, that we found those abnormal fibrils in animal valves – as well as in human valves of all age groups. And what we were dealing with were macroscopically intact valves all down the line. Obviously, the evidence of such bizarre collagenous fibers is invalidated in its pathologic significance by the observations made.

The areas rendered electron-dense by tannin, which occur in particular at subendothelial sites, correspond to the minute – almost punctiform – resorcinfuchsin-positive areas in light-optical staining of elastic fibers. They are to be interpreted as elastin and its microfibrillar preliminary stages ("elastic fiber microfibrils").

4

Functions of Venous Valves

A. Hydrodynamic Function

No further comment is necessary to *W. Harvey's* classic statement that the main function of venous valves is to maintain the direction of the bloodflow to the heart. For that reason the number of valves increases distally so that retrograde waves of blood may be stopped.

In addition to the above-mentioned function venous valves have four differentiated tasks:

1. The valves warrant the function of the peroneal muscle-pump – Fig. 33.

2. The function of the two-phased respirational pump according to Bollinger (1971) – Fig. 34.

3. Synchronization of the impulses exerted by the arteries on the accompanying veins (Hasebrock, 1899) and, thereby, promotion of the bloodflow in central direction.

4. Bloodflow in foot veins. The valves of the perforating veins serve to ensure the bloodflow from the surface into the deep venous system. A major exception is to be seen in the valves of the communicating veins of the foot. Many of these veins have no valves, in which case bidirectional bloodflow is possible. If valves are present, they control the bloodflow from the deep to the superficial venous system, i.e. in the very opposite manner as found in other regions of the leg.

This is due to the fact that with every individual step blood is pressed from the plantar plexus – the French use the term "semelle veineuse" (venous insole) – into both the deep veins and also into the superficial veins. This offers a ready explanation for the fact that the short saphenous vein is disproportionally large as far down as in the ankle region.

One may summarize the function of the venous valve as described hitherto by the term *hydrodynamic* function. It basically consists of ensuring the *direction* of the venous bloodflow to the heart. In addition to the hydrodynamic function, other functions will be discussed briefly below.

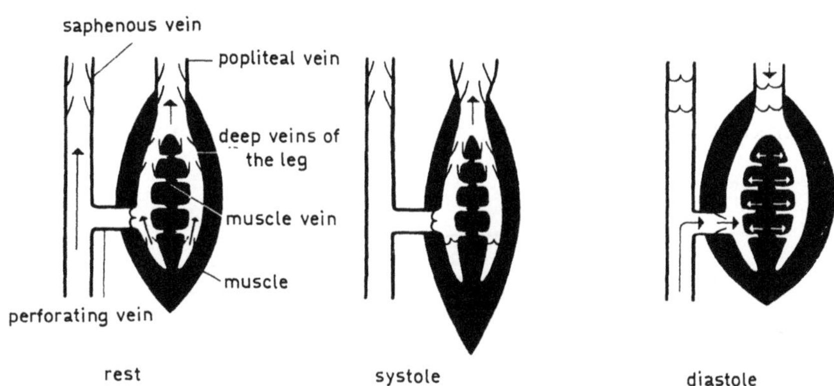

Fig. 33. At rest the bloodflow is maintained slowly and continuously by vis a tergo. All valves of both the superficial and deep venous systems are open, merely the valves of the perforating veins are closed. This is due to slightly differing pressure-conditions prevalent in the deep and superficial veins. The hydrostatic pressure fully affects superficial and deep veins. Muscular contraction = systole. The blood ist pressed in a proximal direction. As the competent valves of the peripheral venous system will promptly close, only one direction of flow is possible. Muscular relaxation = diastole. The pressure in the deep veins of the leg drops rapidly, causing the valves of the popliteal veins to close; simultaneously blood is drawn from the superficial venous system and the muscle veins, resulting in an increase of pressure. This forces the valves of the popliteal veins to reopen – and the cycle to recommence

B. Other Functions of Venous Valves

1. Hydrostatic Function

The term refers to the assumption that sound venous valves damp hydrostatic pressure. Under conditions of erect standing, a blood column of a height of 150 cms would weigh upon the foot veins, unless that pressure were received by venous valves shutting instantly as soon as the central pressure exceeds the peripheral pressure. It might be held against the hypothesis justly that with prolonged standing the bloodflow will not be arrested but maintained (even without any interference of the muscular pump) in a central direction by vis a tergo. During prolonged standing, the valves are open. Therefore, a hydrostatic effect is not to be assumed. – Quite a different situation presents itself under conditions of rapid transition from walkling to standing. During walking, the pressure decreases due to the effect of the muscular pump, and the venous system is largely drained. When the test subject stops, the pressure gradually increases for a period of up to 30 seconds. Subsequently, the pressure in the ankle veins settles down at a level of some 90 mm of mercury.

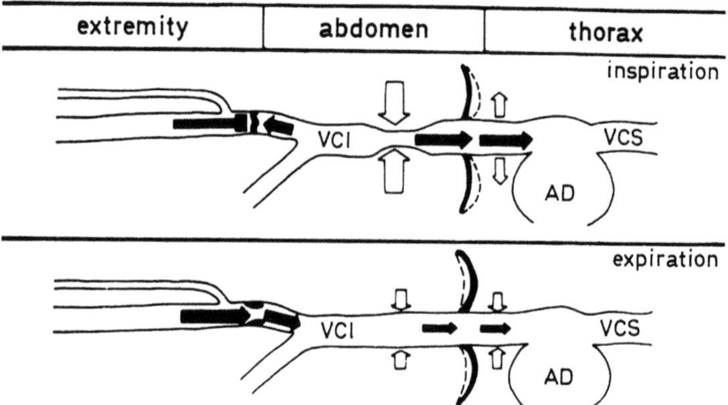

Fig. 34. Thoraco-abdominal suction-pressure-pump according to Bollinger: Schematic representation of the three regions of differing pressure-conditions to be passed through by venous blood on its way from lower extremities to the heart, and the effect of the thoraco-abdominal suction-pressure-pump on the venous hemodynamics in lower extremities. Light-colored arrows = acting ambient pressures of varying intensities; dark-colored arrows = direction and intensity of venous bloodflow; changing positions of diaphragms indicated by bold and dotted lines

This is the moment at which the process of refilling the venous system with blood from the periphery has been completed. We may assume that, until this "steady state" is reached, the venous valves exercise a well-defined function. In the absence of that function the venous system would be filled in a retrograde manner – similar to the superficial veins with *Trendelenburg's* test. Thus, without the effect of venous valves, the maximum pressure would be attained much sooner than we may observe with normal test subjects. – We assume that the venous valves also exercise an essential function during changing of position, e. g. from lying to standing, by preventing major quantities of blood from sinking immediately after the change of position, i. e. from *reflux*.

2. Prevention of Venous Dilation by the Valvular Agger

It has been mentioned before that the vein is particularly thin-walled in the region of the valve-pocket and, therefore, capable of dilatation. On the other hand, the venous wall is thickest in the region of the *valvular aggers,* which reinforce the venous wall and, temporarily at least, prevent the area from dilating, which would cause valvular insufficiency.

3. Hemostasis

In chapter 81 of his novel "Moby Dick", Melville argues that whales will readily bleed to death due to the fact that their veins are completely devoid of valves and that therefore there will be no retarding effect after injuries. – The obvious objection is, of course, that – other than assumed by *Lusitanus* during the 16th century – venous valves do not have any retarding effect at all. Nevertheless, we should not fail to raise the question of whether, under certain circumstances, something like a "Melville-effect" might become operative.

Let us suppose that due to an injury of an extremity a major vein is cut. The injured person loses blood, his blood pressure drops, he may develop a condition of centralization. The blood-supply of the affected extremity is substantially reduced. For this reason, relatively small quantities of the blood flowing back from the periphery are left to be lost. In the absence of more proximal venous valves, he might lose additional blood supplied to the extremity from the caval-vein region in retrograde flow. In a situation like this, sound venous valves will prevent the retrograde bloodflow and, thus, a further loss of blood.

4. Pressure Waves

According to *Ledderhose* (1905), closed venous valves cannot prevent pressure waves from penetrating in distal direction. Pressure waves are caused by coughing, straining, or by rapid manual compression of a venous segment. Due to the elasticity of the venous system, the pressure wave may pass through the valve in peripheral direction. We may assume, however, that in the region of the valve some muting will occur.

5. Fibrinolysis

Using *Todd's* method (1959), *Astrup et al.* compared the contents of tissue-activator of fibrinolysis in venous walls and in the region of the valvular cusps and, by that histochemical method, found a higher content of activator in the valvular cusps.

From a teleological point of view an enhanced fibrinolytic potential in the region of valvular sinuses does make sense, as in that region there is no normal current of blood and, as a consequence, an increased hazard of thrombotic development.

In a more recent work, *Ljungnér* and *Bergquist* (1983) arrived at different results using *Todd's* method. They found some activity of activator in the luminal part of the valvular cusp but otherwise merely activity in the adventitia and no activity at all in the valvular pockets.

6. Conditions of Flow in Valvular Region

It has been mentioned in the chapter on embryology (p. 15) that the flow conditions prevalent in valvular sinuses were studied by *Karino* and *Motomoyia* (1982). Translucent veins were flowed through and the flow-patterns were observed by a cinemicrographic method. Under physiological conditions large paired vortices were proved to be existent in valvular pockets. Small particles entered the valvular sinus and performed a number of successive spiral movements with decreasing radii of curvature. In the course of these movements the particles proceeded to move more and more in a decentral direction until finally they left the valvular pocket. The hematocrit in the valvular pockets was invariably lower than in the main current of the vein. The velocity of flow in the pockets was extremely low so that there was very little shearing-force, allowing the accumulation of cellular aggregates. In valvular sinuses there are marked conditions of hypoxemia, which may even cause endothelial lesions.

By fixation, clearing and dehydrating, the vessels lose some of their elasticity. According to the as yet unpublished work by *Karino* and *Motomoyia* (in press), the valvular cusps will retain their flexibility although they will not completely close under conditions of reversed flow. – Such relative stiffness, however, might result in artefacts of flow processes. Nevertheless, pO_2-measurements taken in live organisms have reaffirmed the findings of *Karino* and *Motomoyia*.

The hypoxemic conditions prevalent in the region of valve pockets were verified by *Hamer et al.* (1981), who took measurements with pO₂-probes. The pO₂-level was extremely low when the conditions of an operation were simulated by immobilization of the extremity for 1 to 2 hours; occasionally it even dropped to zero. – If the extremity was allowed to move freely, the enhanced bloodflow resulted in the formation of eddies and, thereby, in an improved exchange of blood between the valvular pocket and the main current in the lumen of the vein.

7. Breaking Strength Properties of Venous Valves

Venous valves are extremely thin structures, which nonetheless possess astonishing breaking strength properties.

During an experiment conducted with a view to supplying a canine hind extremity by reversed blood flow via the venous system, a vein was cannulized in peripheral direction at the anterior side of the leg at hock-level. Despite utmost efforts, all attempts at manually injecting saline solution were unsuccessful. As we assumed that the tip of the cannula might be located slightly proximal to a venous valve, a fine wire-like probe was inserted into the cannula and used to pierce the tissue in front of the tip. Hereby the resistance was broken – the extremity could be easily perfused by means of the cannula. – This observation serves to illustrate the extraordinary breaking strength of valvular tissue.

If trying to pass through a venous valve in retrograde direction by means of a probe, one might easily find that the wall of the vessel rather than the venous valve will tear.

Doubtless, venous valves derive their remarkable breaking strength from their content of collagen.

According to *Ponomarenko* (1980), the strength of venous valves increases until the 60th year of life and decreases slightly afterwards.

Ackroydt et al. (1985) measured the breaking and tensile strength properties of valvular cusps and adjacent venous walls. The breaking strength of cusps (N per sq.mm) was significantly higher than that of adjacent venous walls. In the region of the valvular sinus the breaking strength of longitudinal strip samples was significantly lower than that of cross strips. If strips were taken from avulvular venous segments, no significant differences between longitudinal and cross strips were observed.

The authors emphasized that the high level of breaking strength inherent to valvular cusps (somewhere between 8 and 10 N per sq.mm) by far exceeds the strength required to withstand the pressures venous valves will be subjected to physiologically. In addition to that, it should be noted that in the region of the valvular sinus the mechanic properties of venous walls are considerably different from those observed in avalvular regions. When the extension was measured at the instant of severance by tearing, the cusps proved to be slightly but significantly more ductile than strips of venous walls.

C. Methods of Testing the Function of Venous Valves

A number of methods is available for testing the function of venous valves. We will first discuss clinical examination and turn to instrumental procedures later on.

a) Clinical Examination

These methods should be familiar to every physician. Their indicative value is relatively high. Clinical examination should serve the purpose of "screening" so as at the very least to sort out those patients that should be subjected to instrumental investigation.

1. Inspection

The presence of extensive valvular damages in the deep venous system may be outruled with a high degree of likelihood under the following conditions:
   – The patient has no swelling in the regions of the ankle or on the distal side of his lower leg;
   – such swelling does not occur in the late afternoon either;
   – the patient is free from any indications of venous incompetence (such as pigmentations, effects of scratching, ulcers or post-ulcerous scars on the distal lower leg).
   (While such symptoms of venous incompetence are reported to be absent in some cases of congenital valvelessness, they will occur more likely than not even in that clinical condition.) (See pp. 81 ff.)

2. Palpation

If, during coughing, turbulences are palpable in the long saphenous vein at the patient's thigh, it is an indication of absence of further proximal competent valves (*Fegan* and *Kline,* 1952). Retrogressive waves in superficial veins are palpable even under percussion of more proximal venous portions (reversed *Schwarz's* test). Positive findings over extended lengths are a clear indication of incompetent valves in the patient's superficial veins.

3. Tests

Finally we would point out *Trendelenburg's* and *Perthes's* tests; the former, if positive, being indicative of valvular incompetence in the saphenous vein, and the latter being affirmative of functional valves in the deep veins of the legs (see pp. 132f.).

By trying to give an exhaustive description of all methods of instrumental investigation available we would go far beyond the scope of this book. Instead, we shall confine ourselves to outlining the fundamentals and suggesting more detailed literature.

b) Doppler-Ultrasound Diagnosis

The question is: are a patient's venous valves functional or not?

Doppler-ultrasound diagnosis offers a simple, non-invasive, repeatable method of investigating into that question.

For further details the monography published by *Kriessmann, Bollinger* and *Keller* (1982) should be consulted.

1. Diagnosis of Valvular Incompetence in Deep Veins

We quote verbatim: "Flow signals read during *Valsalva's* maneuver as well as during compression of the thigh with the probe positioned over the popliteal vein or during compression of the calf with the probe positioned over the posterior fibular vein, are indicative of valvular incompetence between the site of compression and the probe." These signals will not disappear after compression of the superficial veins.

It goes without saying that the flow signals may be recorded by graphs.

2. Diagnosis of Valvular Incompetence in Superficial Veins

Incompetence of Long Saphenous Vein

Incompetence of the terminal valve and the trunk of the long saphenous vein is shown by reflux signals which can be read during *Valsalva's* maneuver. These reflux signals disappear after compression of the superficial veins.

Incompetence of Short Saphenous Vein

Pronounced reflux subsequent to the orthograde signal triggered by compression of peroneal vein.

3. Incompetence of Perforating Veins

We quote from the description of *Partsch's* method given in the monography published by *Kriessmann, Bollinger* and *Keller* (1982):

The patient is sitting with loosely dangling lower legs; the probe is held obliquely in an upward direction. The areas of suspectedly incompetent perforating veins, represented by palpable fascial gaps, have been marked. The draining vein is constricted by means of a rubber-tube applied proximally. Flow signals caused by retrograde bloodflow during contraction of the peroneal muscle or jiggling of the foot are recorded.

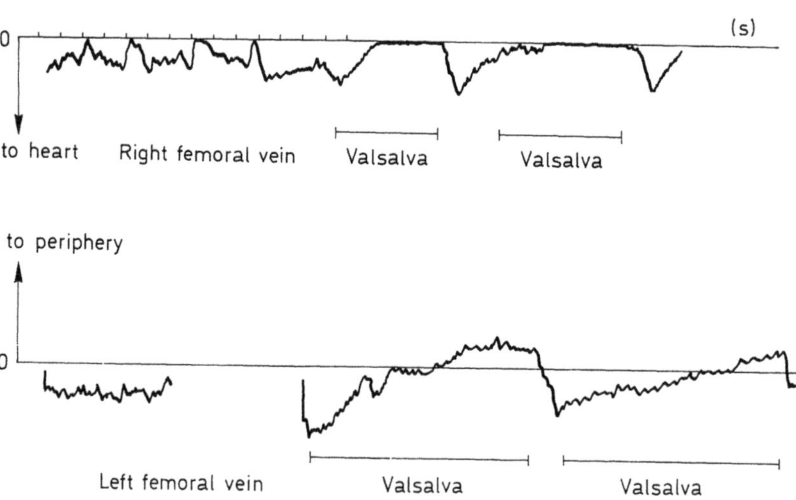

Fig. 35. Sonogram of 59-year old male with post-thrombotic syndrome, left side (status after compound fracture of left tibia). Top: normal findings in right femoral vein, no signals indicative of blood flow in peripheral direction obtained during Valsalva's maneuver Bottom: markedly pathologic result of Valsalva's maneuver, i.e. signals of flow to periphery

Fig. 35 shows the sonogram of a 59-year old male with a status post compound fracture of the left tibia. The sonogram of the left lower extremity was normal. During *Valsalva's* maneuver no signal of flow in peripheral direction was recorded. In contrast, the left femoral vein showed pathology in the form of bloodflow in peripheral direction during *Valsalva's* maneuver.

Fig. 36 shows signals of flow in central direction triggered by compression of thigh and calf and a pathologic signal of flow in peripheral direction obtained during *Valsalva's* maneuver as well as an "overshoot" signal recorded after discontinuation of the maneuver. "Overshoot" is caused by a steep gradient of pressure between the veins of the leg and the central veins. This pressure is built up toward the end of *Valsalva's* maneuver.

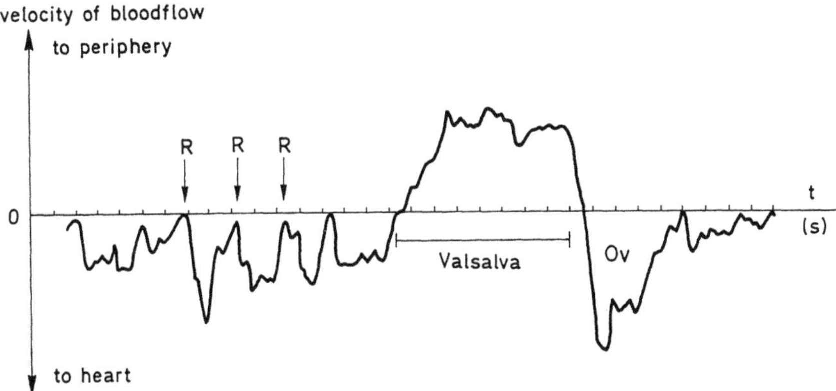

Fig. 36. Bloodflow in region of left femoral vein of 35-year old female white-collar worker with post-thrombotic venous valve lesions involving thigh and pelvic regions (original record) – **A** "A-sounds" triggered by compression of thigh and calf (normal findings with patent deep venous system); **Val** pathologic reflux during Valsalva's maneuver; **Ov** "Overshoot" after Valsalva's maneuver

(By courtesy of the author these graphs were reproduced from Marshall, M.: Praktische Doppler-Sonographie; Berlin-Heidelberg-New York: Springer 1984)

Global competence and/or the extent of incompetence of venous valves may be determined by two methods:

c) Peripheral Venous-Pressure Measurement

(According to *May* and *Kriessmer,* 1978)

In a patient standing upright, the venous pressure in a punctured pedal vein corresponds to the height of a column of water reaching from the site of measurement up to the level of indifference, i.e. the imaginary line, approximately a hand's breadth under the midriff, where the venous pressure remains constant irrespective of the position of the body. If the patient does knee-bends or toe-stands, the pressure will drop imposingly to an invariably constant level, provided that the valves of the patient's calf, popliteal and femoral veins are intact. The greater the extent of valvular lesion, the less noticeable will be the decrease in pressure. This may be recorded in a graph. Depending on the extent of deviation from normal standard readings, defects of valvular function, most of which are caused by post-thrombotic lesions, may be classified according to various degrees of severeness. The method comprehensively involves the entire valvular apparatus of the body and offers an evaluation of the muscle pump (Fig. 37). Normally, the venous pressure should drop to

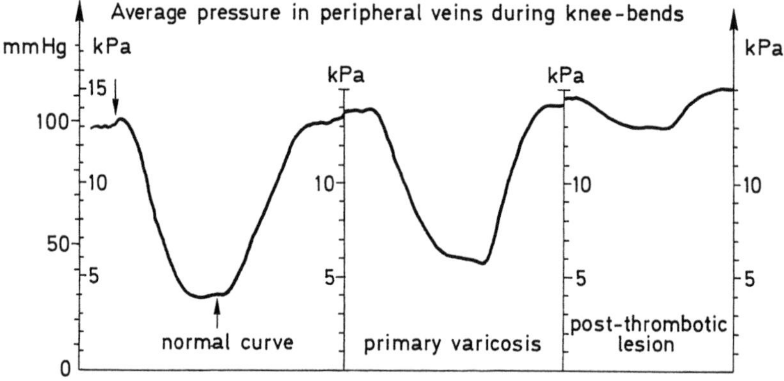

Fig. 37. Venous pressure curves of normal and defective veins

more than 50% of the original level after exercise (*Ferris* and *Kistner,* 1982). Besides the extent to which the venous pressure drops during exercise, there is another aspect of interest, i. e. that of the time the pressure requires to return to its original level after completion of the exercise. Readings of "pressure recovery time" offer information on venous reflux. Readings of less than 20 seconds are considered pathologic.

d) Photoplethysmography (PPG)
and Light-Reflection Rheography (LRR)

The function of the valvular apparatus influences filling and draining of the veins of the dermal plexus. During movement of the foot, the venous plexus empties. When the foot is at rest again, the venous plexus is refilled from the arterial side of the circulatory system. If,

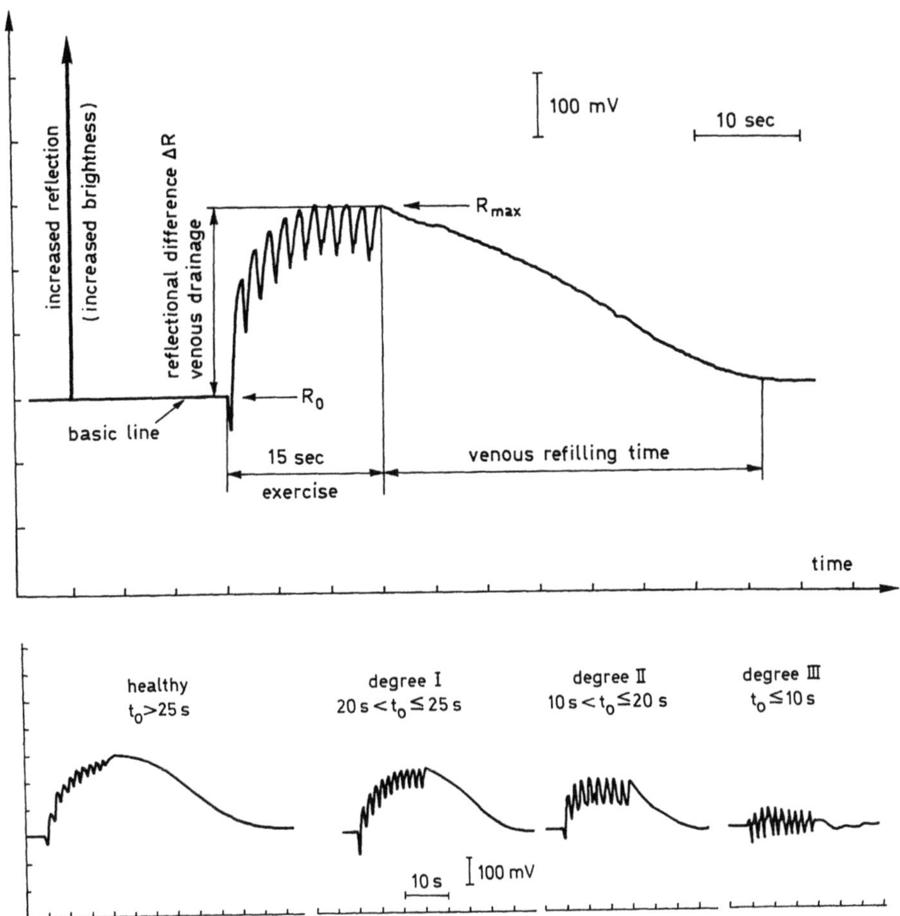

Fig. 38. Upper diagram: normal light reflection rheography. Lower diagram: hemodynamic classification of disturbed venous drainage as determined by LRR

due to valvular incompetence, reflux occurs, such refilling is accelerated.

The reflection of infrared light radiated into the skin may be recorded in a graph.

Photoplethysmography (PPG) (*Barner et al.,* 1978; *Abramowitz et al.,* 1979), a method which has been known for several years, is affected by a number of interfering factors somewhat impairing its value. Light-reflection rheography (LRR), which was developed by *Blazek* (1984) (for details see *May* and *Stemmer,* 1984), successfully avoids these weak points and offers a non-invasive and fairly useful method of indirect examination of the valvular function of leg veins (Fig. 38).

The crucial parameter is that of refilling time. According to *Queral et al.* (1980) the photoplethysmographically determined refilling time largely corresponds to the pressure recovery time in venous pressure measuring. As a matter of fact, the readings of time obtained by the two methods are almost identical (*Kistner* and *Sparkuhl,* 1979; *Johnson et al.,* 1981).

A discussion between *Warren* and *Bergan* and *Yao* (all 1981) has thrown light on one essential difference between the two methods. *Warren* pointed out that often markedly improved photopletysmographic findings were not correlated with corresponding venous pressure readings. In their response *Bergan* and *Yao* stated that the decrease of pressure after exertion was indicative of a functional muscle pump, whereas reduced refilling time was indicative of valvular incompetence at a site more proximal than that of measurement.

e) Other Plethysmographic Methods

For the purpose of supplementing the mainly qualitative information obtained by ultrasound measurement and photoplethysmography, other plethysmographic methods offering defined quantitative information may be used. Due to the complicated procedure involved, the classic method of venous occlusion plethysmography is hardly used for routine examinations. *Thulesius et al.* (1973) and *Norgren* (1974) carried out foot plethysmographies and used this method successfully for accurate hemodynamic analysis of venous diseases.

Lee et al. (1982) investigated various forms of venous diseases by the method of impedance plethysmography developed by *Nyboer* (1970). Still, the simplest method seems to be plethysmography by strain-gauge. The method was first described by *Whitney* (1953). Changes of leg volume are determined by variations of electric resistance of a strain-gauge applied over the circumference of the patient's calf. The gauge consists of a thin mercury-filled tube, the electric resistance of which will increase with extension. Studies by this method were carried out by *Leu* (1969) and *Halstuk et al.* (1984).

Plethysmographic examinations – as well as venous pressure measurements – may be carried out with or without compression of the superficial veins. According to *Aschberg* (1973) plethysmography is the most sensitive method of distinguishing between incompetence of superficial veins and incompetence of deep leg veins. *Schanzer et al.* believe that strain-gauge plethysmography is superior to venous pressure measurement as a method of differentiating between incompetence of the muscle pump and venous reflux.

For more detailed information we recommend the highly instructive review of the subject by *Ehringer* (1979).

f) Funtional Testing of Individual Valves of the Femoral Vein

While up to the present, global evaluation of valvular function was sufficient for clinical purposes, the more recent development of reconstructive surgery of venous valves requires measuring methods of enhanced accuracy and specificity.

1. Retrograde Femoral Phlebography

("Phlebographic Trendelenburg's Test") According to Gullmo (1964)

Due to the fact that injected contrast medium remains in the valvular pocket longer than in the lumen of the vessel, ascending phlebography of patients standing in an oblique position allows fairly accurate evaluation of valves for purposes of routine examinations. The visualization of valves may be improved if the patient strains.

X-ray films offering highest accuracy concerning the functionability of individual venous valves are necessary to ascertain whether valvular reconstruction was successful. These X-rays are achieved by the method of "phlebographic *Trendelenburg's* test" developed by *Gullmo* (1964). We follow his description: "In patients lying in supine position, both normal and pathologic veins are relatively empty of blood so that, when the patient strains, in veins with incompetent valves the entire quantity of contrast medium may pass through the vessel in distal direction without any hindrance by obstacles, and pathologic veins may be filled along the whole leg. If the femoral vein is punctured in the groin, contrast medium is injected, and the patient is told to strain continuously for several seconds, the contrast medium must be stopped completely and intransigently by the first valves it encounters, provided the valves are intact. If, however, the patient strains repeatedly, even healthy valves are passed, and the process of filling only rarely proceeds to a point more distal than midthigh level. Instead, the contrast medium usually drains off in a proximal direction at the first inhalation."

Fig. 39 shows an intact first valve of the femoral vein under straining-type phlebography.

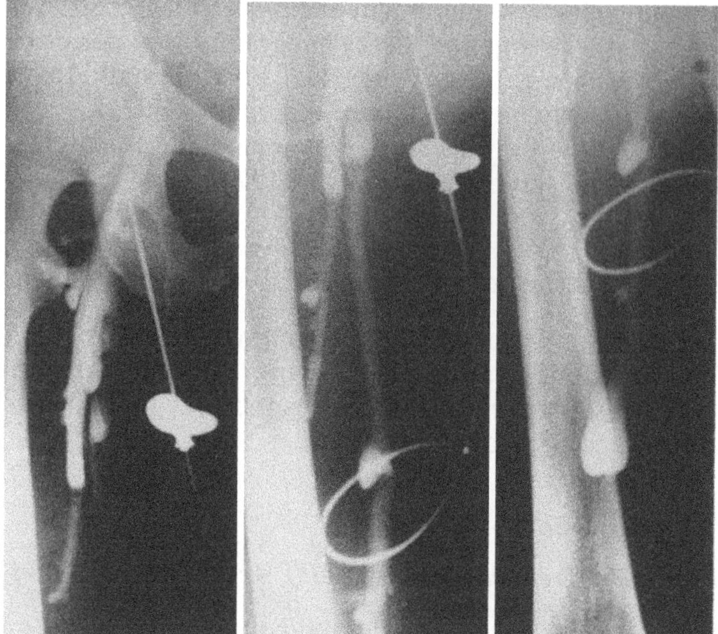

Fig. 39. Competent valves of femoral vein. Visualized by "phlebographic Trendelenburg's test" according to Gullmo after puncture of femoral vein, with the patient in supine position. The left film shows complete stoppage during the patient's first straining. The next x-rays were taken when the patient once more exerted pressure after having taken abreath

For more details on phlebographic diagnosis see chapter on phlebography (pp. 161 ff.).

2. Functional Testing of Individual Venous Valves by Doppler Sonography

This method, which has been improved in recent years, in particular by *Kriessmann* and *Bollinger* (1982), also offers highly specific information on the competence of valves in individual venous regions. Its advantages are non-invasiveness and reproducibility.

For the specific question at stake, we may obtain three significant curves:

1. normal venous blood flow in femoral vein (Fig. 35 – upper left side),

2. blood flow during *Valsalva's* maneuver with intact valves (Fig. 35 – upper right side), and

3. retrograde blood flow during *Valsalva's* maneuver with incompetent valves (Fig. 35 – bottom).

Summary of Part II

Occurrence and macroscopic anatomy of venous valves have been described. In the extremities the cross-section of veins is elliptic, the free borders of the valvular cusps are located in the longitudinal axis of the ellipse and aligned parallel to the skin and the muscular fascia. With advancing age, valves are liable to regress in varying degrees, depending on regional factors. In addition to conventional histologic sections, semi-thin sections and silver stained en-face preparations are efficient methods of visualizing venous valves. Electron-optical studies revealed sporadic occurrence of smooth muscle cells in the entire valvular region. Myofibroblasts, occurring in increased quantities at the edge of valvular cusps, represent the majority of connective tissue cells found. The collagenous fibrils are considerably irregular. Elastic fibers are rarer than one would expect under light-optical microscopy. – The main function of venous valves is to ensure unidirectional blood flow, assist the muscle pump and prevent reflux. Methods of functional testing are discussed (clinical examination, Doppler-ultrasound method, venous pressure measurement, light-reflection rheography [photoplethysmography], and angiography).

Part III. Pathologic Venous Valves

1

Pathologic Morphology

A. Mechanical Lesions, Chemical Noxae

In normal veins mechanical traumatization or perfusion of highly acidic, alcaline, hypertonic or hypotonic solutions or protein-binding substances results in largely similar morphologic changes of the endothelium. Endothelial cells scale off singularly or in groups or, in the case of more intensive noxae, completely. Instead of the pattern of endothelial cement-lines predominantly aligned with the longitudinal axis of the vessel, "transverse lines", i.e. silver lines visualizing the deeper layer of the media and aligned transversely to the axis of the vessel, can be found in increasing quantities. It is only certain noxae, such as cationic detergents, free iodide-ions or organic solvents that completely prevent cement and transverse lines from being stained.

In pathologic venous valves a different picture presents itself. Due to the fact that in valvular cusps there is no layer of smooth muscles, whose cells are surrounded by "transverse lines", such lines do not occur in pathologic cusps. In silver stained specimens of chemically or mechanically damaged venous valves silver lines are absent focally or over the whole area. On its surface, the valve visualized in Fig. 40 A shows small spindle-shaped spots of darker visualization which might correspond to detached or dissociated endothelial cells. In contrast to that, typical transverse lines are visible in the endothelium of the venous wall situated below (Fig. 40 B).

B. Congenital Absence of Valves

While, presumably, congenital absence of valves in deep veins was first described by *Luke* (1941), it was not before the publication of the monography by *Lindvall* and *Lodin* (1961) that widespread attention

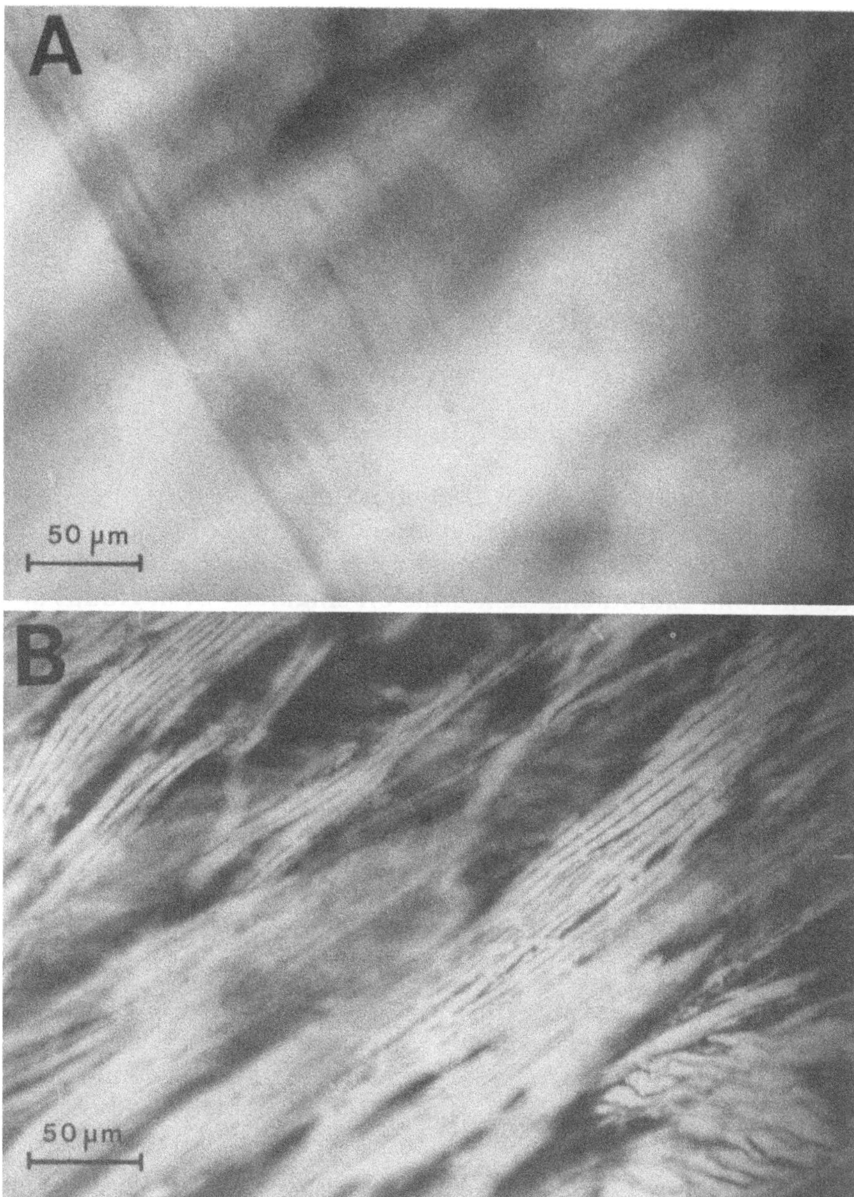

Fig. 40. Venous valve from canine jugular vein. Damage caused by flushing with tris-buffer, pH 11.8. Lower picture: **B** Parietal endothelium with transverse lines and endothelial insula in lower right corner. **A** Same area shown under raised focus of microscope. Level of valve. On the valve no silver lines but merely darkened oblong spots are visualized – presumably dissociated and shrunk endothelial cells

was given to that clinical picture. Meanwhile, reports have been forthcoming in increasing numbers. While we are in no position to give any percentage with respect to the frequency of occurrence of the syndrome, we can state that year after year several cases have been observed.

In theory it would be impossible to differentiate by means of a single X-ray film between congenital avalvulia and age-related destruction of valves caused by thrombosis of valvular pockets. The combination between characteristic clinical picture and objective findings is so unambiguous, however, that hardly any doubts ever arise in differential diagnosis.

1. Symptoms

The case history outlined below is typical and in correspondence with the cases reported by *Lindvall* and *Lodin* (1961), *Bollinger* (1971) and *Hepp* (1980).

J. M. was transferred to us with a diagnosis of thrombosis. On admission her legs were clinically normal. A phlebographic examination revealed that the deep veins of both her legs were completely avalvular apart from one valvular rudiment. The tonus of the veins was normal (Fig. 41).

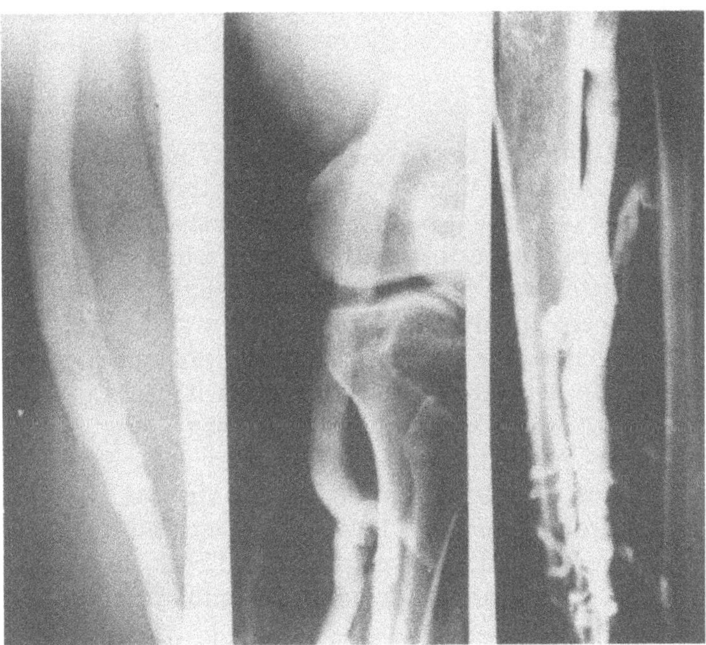

Fig. 41. Bilateral congenital avalvulia. Right leg. The veins of the left leg show identical findings

Past history: The patient had been suffering from swollen legs during summers since the age of 16. Two years before admission to our clinic, she had been hospitalized for bilateral thrombosis. At that time the "thrombosis" was healed by heparin within a few days. Previous to her being transfered to us she was given fibrinolytic treatment and there was similar recovery. A phlebographic examination had not been made up to this time.

The past histories of all such cases are strikingly similar:
1. First occurrence of symptoms between the ages of 11 and 16; disappearance of edemata overnight;
2. equal involvement of *both* legs;
3. in some cases there are florid or healed crural ulcers, sometimes on both legs.

It is not surprising that many variations ranging from reduced valves to complete valvelessness can be found. Nor is it surprising that occasionally there may be other anomalies, e.g. teleangiectatic nevi, subcutaneous hemangiomata etc., as described by *Hepp.*

Bollinger (1971) thoroughly investigated 19 patients with congenital avalvulia. He also described past histories of uniform appearance.

Expectedly, Doppler-ultrasound examination revealed an absence of flow-arrest during Valsalva's maneuver, and continuous flow instead of a respirational phase during deep respiration. In phlebography some singular valves were occasionally found in the veins of the calf, whereas the valves of the muscle veins were usually present.

2. Heredity

Lindvall and *Lodin* (1961) proved heredity of avalvulia over several generations in some of their cases. Even if there are no exact case histories for the less recent past, anamnestic statements according to which the patient's ancestors had suffered from swollen legs since youth, that the edemata had subsided over night, that some of them had crural ulcers involving both legs under the age of 20, are conclusive evidence. The hereditary factor is not surprising in view of the fact that the heredity of varicose veins is generally held to be between 70 and 85%. A thorough study involving 15 patients of 3 generations of one family was published by *Plate et al.* (1983). The authors comment about the heredity of the disease as follows:

"Only descendents of individuals with a verified or probable vein valve aplasia were noted to have aplasia and both sexes were equally affected. This makes an autosomally dominant inheritance most plausible.

Even if the results of active treatment are limited, the establishment of this diagnosis is valuable to the patient since genetic information is possible. The dubious long term prognosis is also important in the

selection of an occupation and carreer for a young individual.
Standstill employment should be avoided."

3. Pathophysiology

The causal genesis of sequelae as described by *Lodin* and *Zetterquist*
is summarized in the scheme given in Fig. 42. The sequelae of
congenital valvelessness are largely similar to those of age-related
loss of valves.

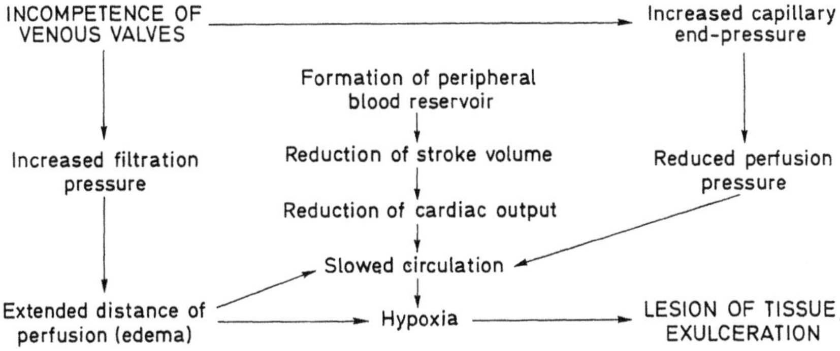

Fig. 42. Mechanisms that may lead to lesions of tissue and lower-leg
ulcerations under conditions of incompetence of venous valves (Lodin and
Zetterquist)

C. Age-related Changes

a) In Conventional Paraffin Sections

Age-related changes in venous valves were described by *Saphir* and
Lev (1952 B).

In the *parietal portion* of venous valves the delicate areolar tissue is
replaced by thick and dense-collagenous fibers. These changes
commence at about the age of 30. The crypts become flatter and less
noticeable. Starting at the age of 40, the elastic fibers forming an
extension of the elastic membrane in the luminal part increase in both
quantity and thickness. Under the age of 40 such fibers are sparse.

In the *luminal part* deposits of connective tissue of minor
distinctness are found between endothelium and elastic membrane
until the 10th year of life. From the age of 10 their distinctness
decreases until the age of 40, at which time it increases again. From
that age onwards also endophlebohypertrophy of the vein itself,
proceeding into the proximal third of the valve, is observed.

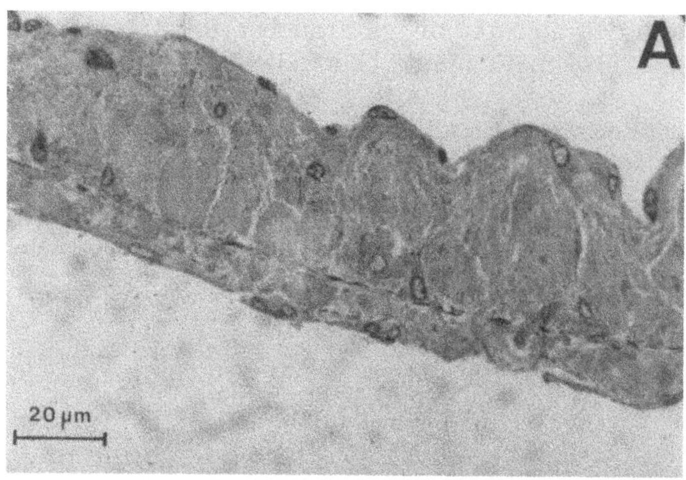

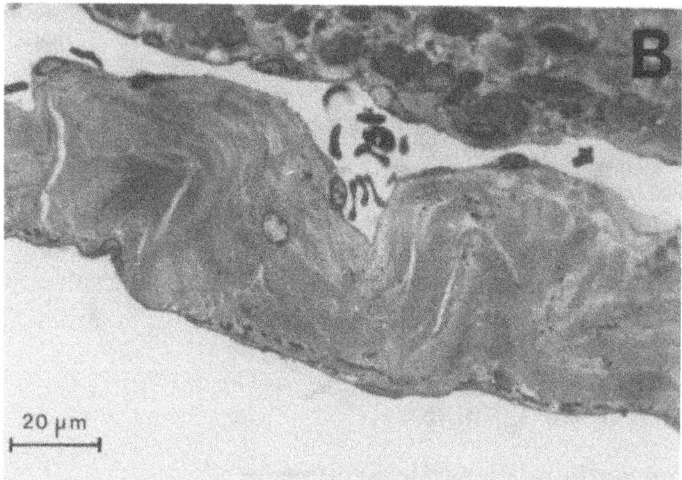

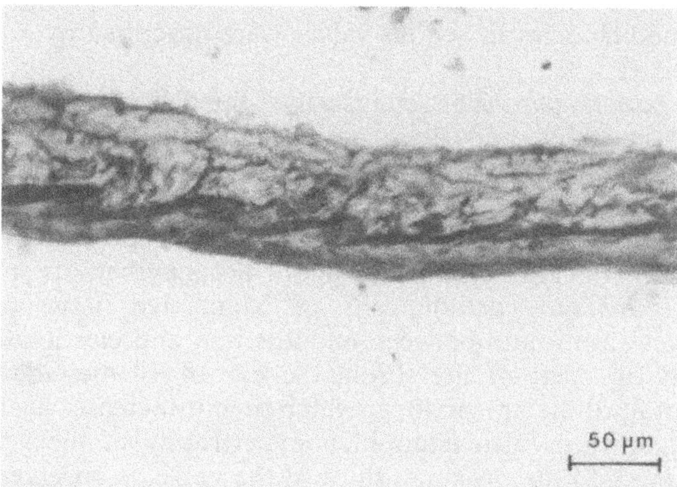

In young individuals the elastic membrane is almost straight-lined. In old individuals there are indentations resembling "fjords in a jagged coastline". New formation of thin elastic lamellae begins between the 30th and 40th years of life.

The findings of *Saphir* and *Lev* have been fully confirmed with respect to both varying thickness of the intima in young, medium-aged and old persons, and augmentation and jaggedness of the elastic tissue in age (Fig. 43 A–C).

In a venous valve of a 31-year old patient, *Saphir* and *Lev* found a cyst-like structure, which contained erythrocytes and was surrounded by collagenous fibers. The cyst was situated between elastic membrane and endothelium, projecting into the lumen of the vessel. It was comparable to blood cysts, occasionally found in cardiac valves of newborns.

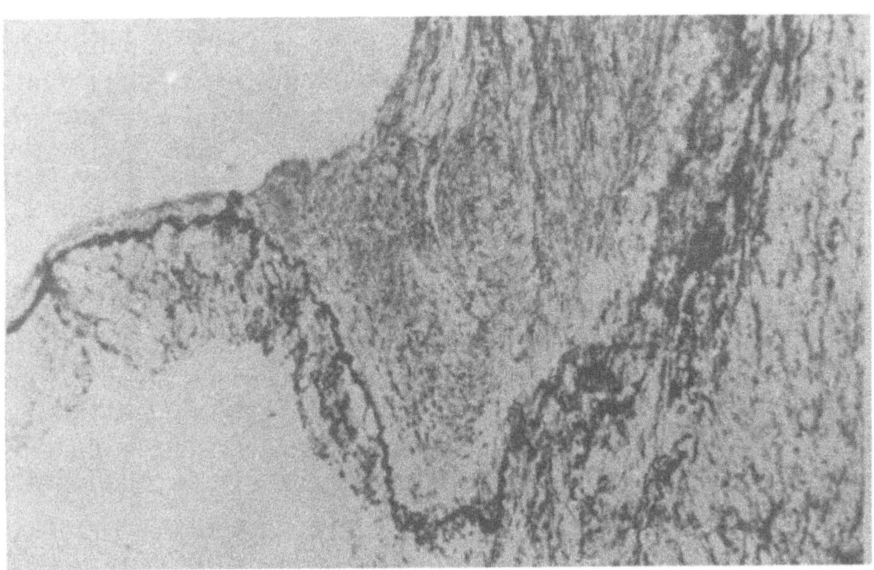

Fig. 44. Age-related changes: "increased endophlebo-hypertrophy" below valve, proceeding into luminal part of valve. Magnified approx. 150 times. [From: Saphir and Lev, Amer. Heart J. *44* (1952)]

Fig. 43. States of intimal layer at different ages. **A** Relatively wide intimal layer in 19-month old infant. **B** Very narrow intimal layer in 37-year old male. **C** Widened intimal layer and increase of elastic elements in 70-year old female. **A** and **B** Semi-thin sections, **C** paraffin section, orcein-hemalum stained

Deposits of fat have never been found in venous valves. Even in connection with age-related changes, in-growth of vessels into venous valves has *not* been observed.

The signs of age in venous valves are different from the picture of healed valvulitis. There never is augmentation of cells in old veins, whereas "fjords" in the elastic membrane were never found in valvulitis cases.

Endophlebohypertrophy (Fig. 44) is found mainly distal from venous valves. *Saphir* and *Lev* discuss whether the reason for this may be that if the valve is even slightly rigid, eddies might form distally from the valve, which would secondarily cause hypertrophy of the subendothelial connective tissue. The explanation does not seem satisfactory at first though. After all, one would assume that eddies will form on the downstream side of an obstacle rather than on the upstream side. On the other hand, the hypothesis put forward by *Saphir* and *Lev* finds reaffirmation in an observation *Cotton* (1961) was the most recent investigator to make: as a rule, aneurysmatic dilations of varicose veins are also found mainly *distal* from valvular cusps.

b) Age-caused Changes in Endothelium

(Monolayered Flat-Preparations)

The results of age-caused changes are manifested in the endothelium by the occurrence of large polynuclear cells. So far we have found such spontaneous age-caused changes only in humans. They have not occurred in the animals used in our experiments, presumably because those animals never reach an age at which changes occur. In humans, however, these age-caused changes are very distinctive (Figs. 45–47).

In laboratory animals such age-caused changes may be simulated by traumatization of vessels. *Gottlob* and *Zinner* (1962) distinguish between "hard traumas" of severe nature (the vessel is squeezed by a normal hemostat) and "soft traumas" of mild nature (the vessel is squeezed gently by a clamp, the branches of which are covered by rubber tubes). Hard traumas cause circumscribed detachment of the endothelium. The traumatized region will be overgrown with new endothelial cells. Giant cells rarely occur. In the case of "soft traumas" the cells do not perish immediately but necrotize gradually and are subsequently replaced by new cells. In this process multinuclear giant cells do appear in considerable quantities. – It is to be assumed that aging of man is comparable to protracted traumatization so that in aging man giant cells occur in large quantities.

Salomonowitz and *Gottlob* (1981) reported on a quantitative study of the occurrence of giant cells in the common iliac vein. They found that in the left vein giant cells occur more frequently than in the right one. They attributed this difference to the fact that the right common iliac artery passes over the left common iliac vein in transverse direction and presses it toward the promontory. They argued that this was a combination of the factors of aging and "soft trauma", the latter being caused by pulsation of the artery. – Moreover, in this particular region there are spur-like excrescences which may have a constricting effect upon the lumen of the left common iliac vein (*May* and *Thurner,* 1956).

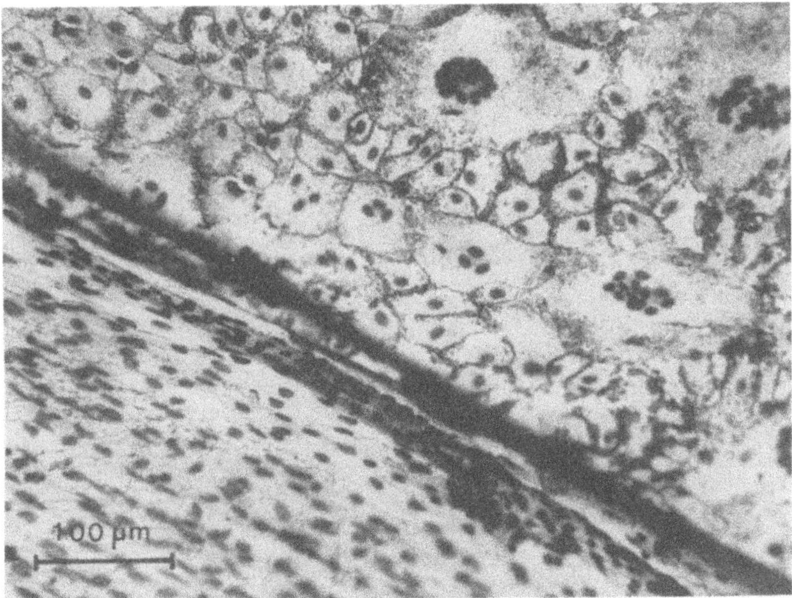

Fig. 45. Saphenous vein, 50-year old male. Borderline between venous wall (upper right side) and valvular cusp (lower left side). Monolayered ("Häutchen") preparation

Investigations of human leg veins reveal giant cells of increasing quantities starting at the age of 50. In venous valves, however, they are rather rare findings. There is a clear difference between valvular endothelium and the endothelium of other parts of veins (Figs. 45–47).

While binuclear cells were found in valvular endothelium, cells having more than two nuclei were not detected. Giant cells were rare (Fig. 47 B).

That age-caused changes in the region of valves are considerably less frequent than in the adjacent venous wall is indeed a surprising

 Pathologic Morphology

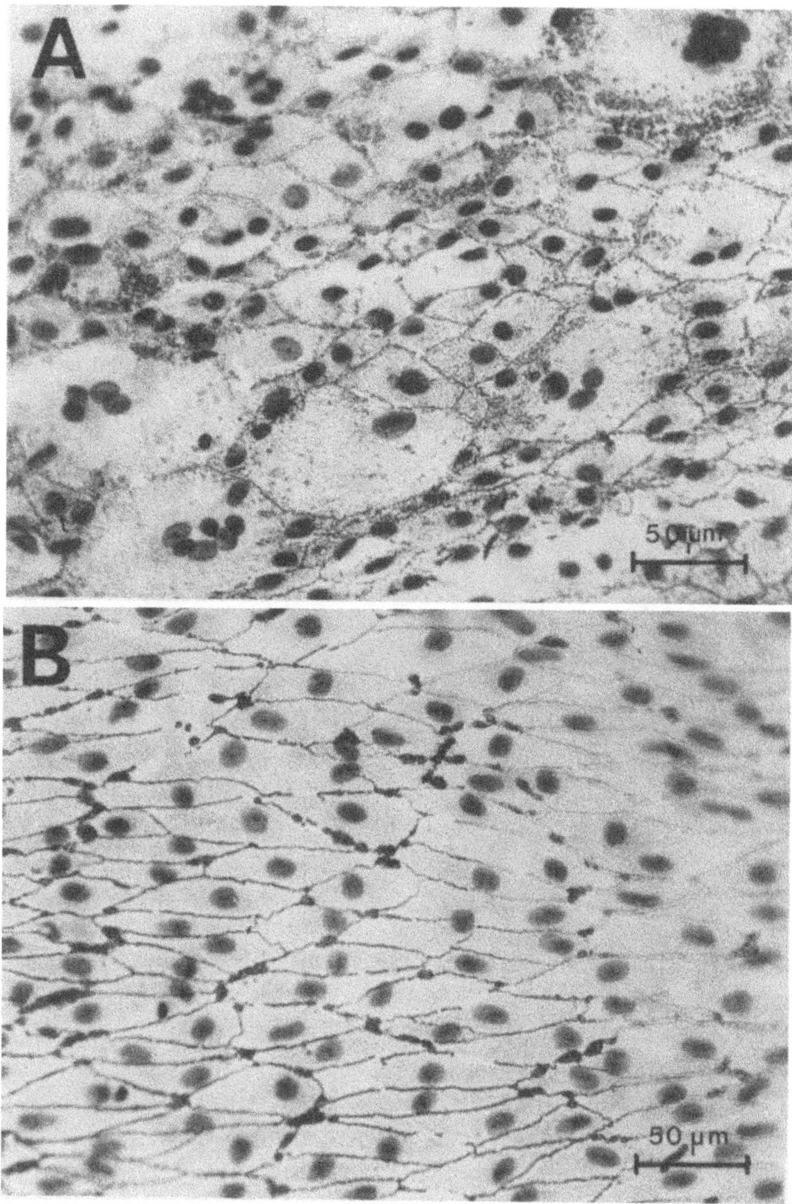

Fig. 46. Human vein, unicellular flat ("Häutchen") preparations. **A** Venous wall with numerous giant cells, predominantly polynuclear. **B** No giant cells in adjacent valvular cusps

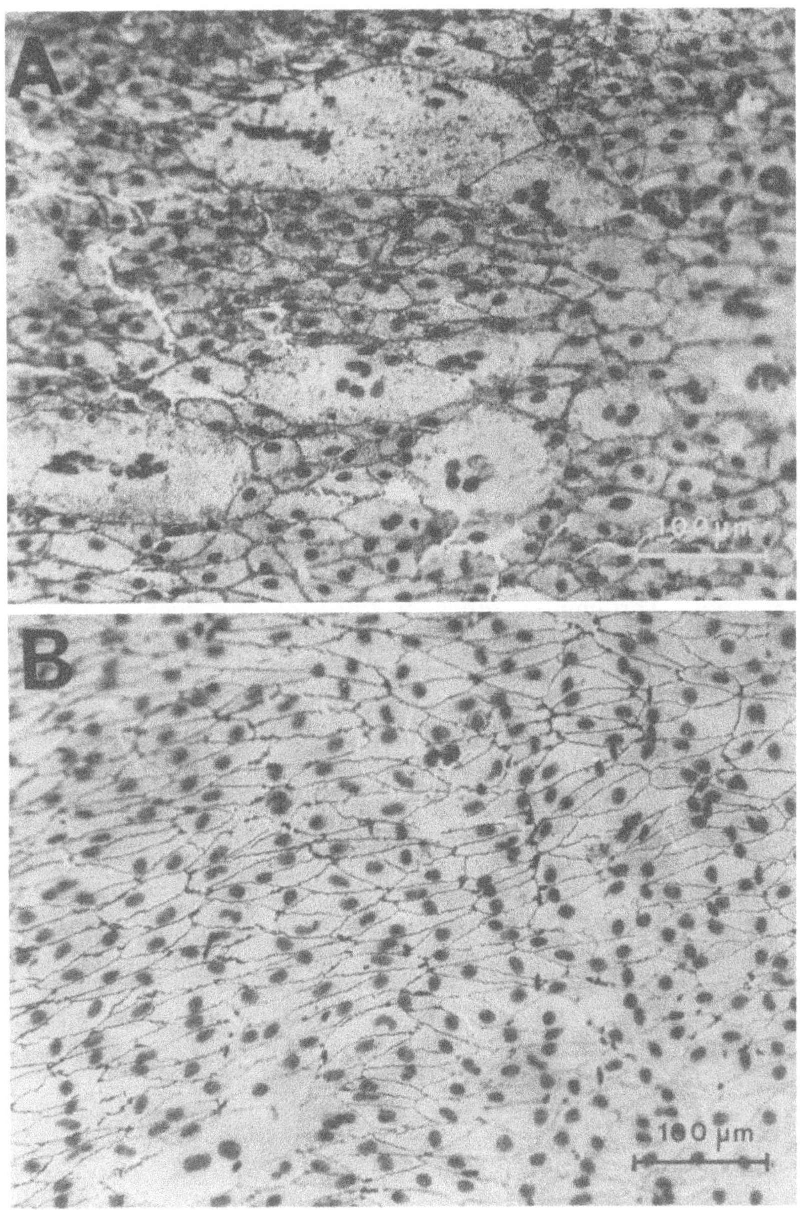

Fig. 47. Femoral vein of 70-year old male, unicellular flat ("Häutchen")
preparations. **A** Venous wall with numerous polynuclear giant cells.
B Valvular cusp with only one giant cell but several binuclear endothelial
cells of normal size and manifestations of cell division

finding. After all, one would tend to assume that, moving much more than the venous wall, valves should show such changes sooner than the venous wall. On the other hand, the relatively frequent occurrence of cells with two nuclei (instead of greatly enlarged cells) might be explained by the assumption that in the valvular region there is more frequent regeneration of the endothelium by cell division than in the venous wall.

D. Inflammation of Venous Valves

Venous valvulitis was described by *Saphir* and *Lev* (1952). They studied venous valves of 25 patients suffering from endocarditis. Cases of both acute endocarditis and chronic rheumatoid disease were included. – In five cases changes of peripheral venous valves were found.

Case 1: Acute bacterial endocarditis superimposed on old endocarditis of the aortic, mitral and tricuspid valves. In the right femoral vein a mural thrombus was adherent to a valve. The cusps of that valve were necrotic in some portions. Throughout other portions of the cusps were many polymorphonuclear leucocytes. The lace-like collagen of the parietalis was replaced by young connective tissue which was rich in nuclei and which contained a number of young blood vessels.

Case 2: 28-year old woman, abacterial thrombotic endocarditis of the mitral valve superimposed on old endocarditis. Thrombosis of the left femoral vein, the thrombus attached to the intima just beneath the valve. The valve grossly normal. Microscopically: a number of polymorphonuclear leucocytes in the subintimal layer of the wall of the vein. Many inflammatory cells through the entire cusp. In some fields circumscribed accumulations of these cells, intermingled with red blood corpuscles were attached to the luminalis. These accumulations resembled vegetations.

Case 3: 6-year old boy, acute rheumatic pancarditis and old endocarditis. Femoral veins grossly normal. In one femoral vein close to the agger of a valve moderate infiltration of polymorphonuclear leucocytes and lymphocytes. Such cells in foci were also found in the parietalis.

Case 4: 29-year old women, acute episode superimposed on chronic rheumatic endocarditis. The femoral veins were grossly normal. On microscopic examination the normal structure of a valve was preserved. There were, however, a number of polymorphonuclear leucocytes in the midportion of a cusp. Focally, an increase of the connective tissue of the parietalis was observed. There was a multiplication of the lining cells of the luminalis in certain foci.

Case 5: 63-year old women, endocarditis of the mitral valve. No gross changes in femoral vein valve. Microscopic examination revealed depression of the commissural mound actually below the level of the adjacent wall of the vein. The region showed absence of muscle fibers and masses of thick

elastic lamellae but no inflammatory cells. Endophlebohypertrophy immediately distal to the valve. In the cusps reduplication of elastic lamellae of
the luminalis. In the parietalis an excess of coarse connective tissue. To the
free end of the cusp a small apparently old thrombus was attached. There
were no blood vessels in the cusp.

The investigations by *Saphir* and *Lev* clearly showed that during
the course of endocarditis there may be inflammatory changes in the
regions of the cusps, whereas other changes are more likely to be
identical with more recent or past episodes of thrombosis of the
valvular pocket, which are described below.

E. Venous Valves and the Development of Thrombosis

Paterson and *McLachlin* (1954) very carefully dissected femoral veins
and their tributaries *in toto* from 165 human cadavers and found
incipient thrombi in 21 cases. Serial sections of these thrombi were
prepared in order to find out at what particular sites these thrombi
had developed. In addition, the authors removed veins from 100 human cadavers and prepared sections for the purpose of comparing
wall lesions in thrombosed and non-thrombosed cases. They found
that phlebosclerosis and other changes were no less marked in the
control cases than in the cases with thrombosis. Even in the
thrombotic cases the primary clots usually occurred in regions of
minor changes of the venous wall. Closer examination of the
21 vessels with incipient thrombi by means of serial sections revealed
that in 17 cases the origin of thrombosis was to be found in valvular
pockets. Frequently, no other than parietal valves were affected
(Fig. 48). The authors assume that, because of the absence of any
substantial current of blood in valvular sinuses, they were sites of
predilection for the incipience of thrombosis. The primary thrombi in
the valvular sinuses often were accompanied by appositional thrombi
constricting the lumen of the vessel or completely occluding it over
distances of varying length.

F. Thrombosis of Valvular Pockets (Figs. 48–50)

Other authors reporting on thrombosis of valvular pockets are *Gibbs*
(1954) and *Sevitt* (1974, 1974 a). *Sevitt* found valve-pocket thrombosis
in femoral veins of cadavers. He describes a total of 48 cases of
valve-pocket thrombosis in femoral veins of 39 individuals. In a large
majority of cases, the thrombus was adherent to the parietal
endothelium, whereas the valve itself was in adhesion with the

thrombus in only a few cases, and the cusp was the site of origin of the thrombus only in exceptional cases. The cusps were frequently completely unchanged but occasionally thickened. In some cases "double thrombi" were found, i.e. recent thrombotic deposits on sites of previous thrombosis.

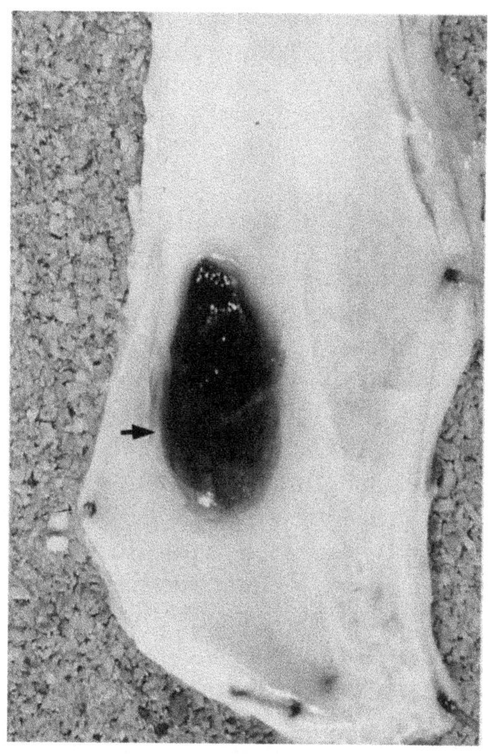

Fig. 48. Valve-pocket thrombosis. Femoral vein of 70-year old male. Macroscopy, Arrow: Susertion of the translucent cusp.

Karino and *Motomiya* (1982) argued that the extremely low velocity of flow in valvular pockets, along with the formation of eddies and the enhanced tendency toward cellular aggregation in this region, offer highly favorable conditions for the occurrence of thromboses in valvular pockets. This is particularly the case since this region is also the site of endothelial lesion caused by hypoxemia, as a consequence of which blood elements become adherent to the venous wall.

McLachlin et al. (1960) investigated into the question of how long it takes until injected radiopaque contrast medium is washed out from the veins of an extremity. With the patient in supine position, similar to the conditions prevalent during surgery, there was evidence of contrast medium being retained in the valvular sinuses for as long as 27 minutes after the injection in some cases. The authors believe that these flow patterns of low velocity are an explanation for the observation that valvular sinuses are regions of predilection for thrombotic development. *Paterson* and *McLachlin* (1954), *Lund et al.* (1969), and *Diener* (1971) also believe that venous thrombosis has its

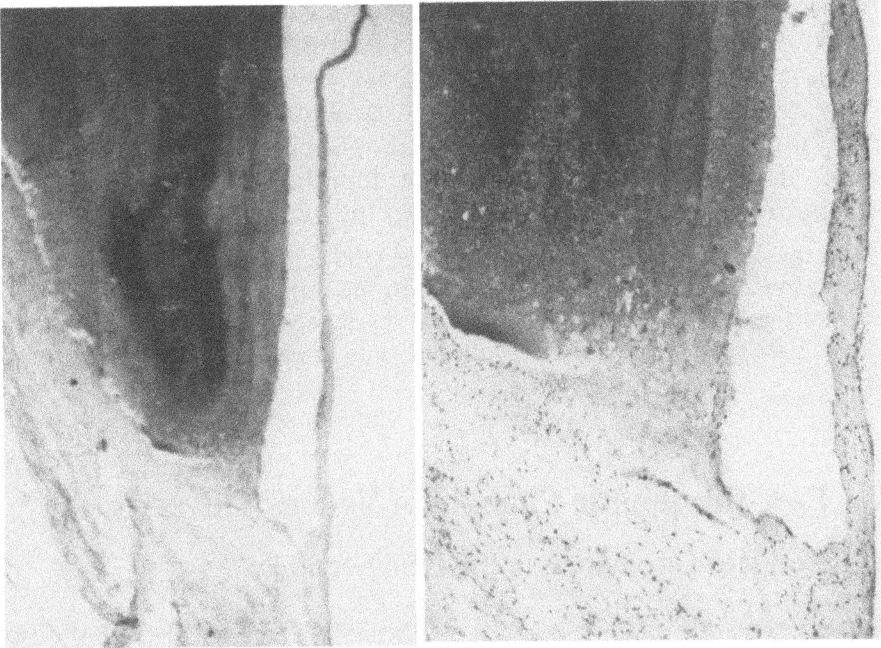

Fig. 49. Valve-pocket thrombosis in femoral vein of 68-year old male. Left: low power magnification, magnified 25 times. Right: site of attachment of the thrombus, magnified 62.5 times. The valvular cusp is unaffected. Typically, the thrombus adheres to the venous wall in the region of the valvular sinus. Signs of incipient organization

origin in the region of valvular pockets. In incipient, non-obturating thrombi the oldest portion of the thrombus, showing manifestations of organization, is found at the lowest point of the valvular pocket. When growing until finally occluding the vessel, the thrombus grows by apposition in proximal and distal direction. Such growth is favored by the condition of stasis, which meanwhile has developed in the vessel. When the thrombus has reached that stage, its place of origin in the valvular pocket is obfuscated. *Lund et al.* (1969) state that the frequency of thrombotic occurrence in a given venous segment is in correlation with the respective number of valves. In valveless veins, such as the caval vein or the common iliac vein, and sparsely valved veins, such as the external iliac vein or the popliteal vein, thrombi occur rarely, whereas multi-valved veins, such as the femoral vein or the veins of the lower leg, show a high level of thrombotic incidence. According to *Lodin* (quoted after *Lund et al.,* 1969), the frequency of thrombotic development is said to be low – despite the impaired flow patterns – in patients with congenital avalvulia. *Diener* (1969) reports on extensive studies of cadaver material, which was also subjected to postmortem phlebography. In 300 cases, 46 valve-pocket thrombi of varying sizes were found. Whenever a thrombus was attached to a valvular pocket by one of its ends, it did not cause venous occlusion. Such thrombi are clinically mute. Thrombi of less than 2 cm length, with extremely rare exceptions, were found to be inserted into valvular pockets.

On the other hand, according to *Havig* (1977), the number of valves in a femoral vein is in inverse correlation with the number of thrombi found in that particular vein. *Havig* assumes that, by receiving hydrostatic pressure, valves will protect the venous wall and thus prevent thrombotic development.

G. Post-thrombotic Changes of Venous Valves

a) Complete Thrombosis in Paraffin Sections

Edwards and *Edwards* (1937) thrombotized femoral veins in 16 dogs by injection of sclerosing solution (Na-Morrhuate). After intervals of varying duration, the veins were dissected and subjected to histologic examination. The following findings were made:

Complete thrombosis results in valveless veins. If, after thrombotization, the cusp is in the center of the vessel in a closed position of the valve, it is affected by the process of thrombotic organization. It is fragmented, occasionally adherent to folds of its own or of the opposite cusp, and finally it is infiltrated by fibroblasts and capillaries. While, in the course of reorganization and recanalization, the capillary structures increase in proportion, the valvular cusp gradually disappears, so that there will be no more than occasional rudiments of its elastic membrane in trabeculae, which are projecting out into the new lumen. During a later phase these rudiments of the elastic membrane are also found in a newly formed, irregularly thickened intimal layer.

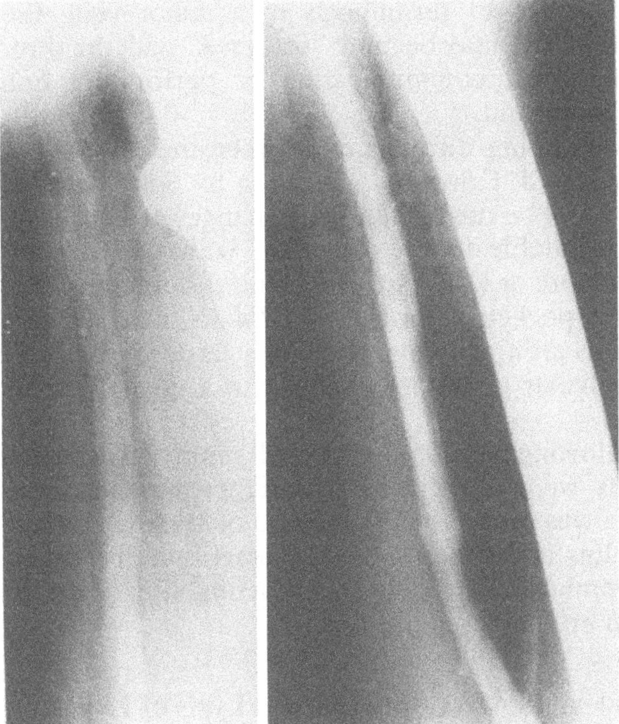

Fig. 50. The valves of the femoral vein are no more than rudiments. In view of the fact that the counterlateral vein shows perfectly normal valves, it is to be assumed that the absence of valves is due to valve-pocket thrombosis of unnoticed course

Cusps located in the proximity of the venous wall after thrombotization are not destroyed but usually "soldered" to the venous wall so that, once again, the vessel is left valveless.

b) Incomplete Thrombosis

Incomplete thrombosis may cause stenosis and incompetence in the same manner as it is observed with cardiac valves. Fragmentation may be of minor nature and restricted to the base. Proximal portions of the cusp may be fixed to the venous wall by new connective tissue. The valve is thickened and shortened either over its entire dimensions or only in certain portions. Stenosis may occur if, in the valvular pocket, there is formation of thrombotic cushions constricting the lumen. Occasionally complete adhesion of the valvular cusp to the venous wall is observed.

In the presence of thrombosis in a major vein, the valves of entering tributaries may become "soldered" with the thrombus in the major vein. Thus, communicating or perforating branches may become incompetent.

There are striking differences between the observations of *Sevitt* and *Edwards* and *Edwards*. According to *Sevitt,* the valvular cusp frequently is not be subjected to any changes at all. These differences may be attributable to the fact that, while *Sevitt* studied natural thrombosis more or less confined to the endothelium of the vascular wall in valve pockets, *Edwards* and *Edwards* caused primary endothelial lesions arbitrarily by means of sclerosing fluids. Understandably, the valvular cusps are changed to a greater extent by such a procedure.

In the following paragraphs we shall report on experiments during which veins were obliterated without primary endothelial lesion. Thrombosis was caused by the method of *Wessler et al.* (1955), which produces clots and changes of the vessel that are largely similar to clinical thrombosis. The changes occurring after recanalization were observed in en-face preparations.

c) Venous Valves After Spontaneous Lysis of Experimental Thrombi (En-face Preparations)

Studies (*Gottlob* and *Kimmel,* 1973) were undertaken with a view to ascertaining what morphologic changes are caused to venous valves by thrombotic development. Investigations involved the anterior and posterior facial veins of rabbits. These veins unite in the neck region, forming the external jugular vein, a relatively large vessel, which – together with its counterlateral partner-vein – drains almost all of the blood supplied to the head. In these veins thromboses were produced by means of a modified version of *Wessler's* method (*Wessler et al.,* 1955).

After paramedian incision of the skin under anaesthesia obtained by barbiturate, the vein of one side was exposed just far enough that its course was clearly visible. The vessels were transfixed by surgical needle and synthetic suture above the junction of the facial veins and at the trunk below the junction. The sutures were passed through the skin to the outer side of the body. 5 ml of homologous serum was injected into a collateral ear vein. After 5 minutes the sutures were tightened so as to constrict the vessel at sites proximal and distal to the venous valves. The sutures were tied around a rubber tube fastened to the outer skin in such a manner that they could be easily removed after 48 hous. At some time between 5 and 111 days after surgery the animals were anaesthesized once more, and the thrombotized vessels were removed and (in the manner that has been described elsewhere in this book, pp. 42 ff.) spread on cork plates, cut open, stained by silver nitrate and processed to whole-thickness en-face preparations. It was found that, at a point of time approximately 20 days after the operation, the venous segments either had been thrombotized or affected by valve-pocket thrombosis. Between the 5th and 20th days after the operation there were occasional findings of normal valves, or at least of normal appearance under light-optical microscopy, but in the majority of valves there were marked changes, the intensity of which was more enhanced in older specimens than in those taken during the first weeks. While on the 5th day there still were large-sized thrombotic masses in the vessels, thrombi as a rule were not found from the 9th day onward. Instead, there were parietal plaques. In silver-stained specimens the following changes were observed:

1. Thrombotic plaques on the valvular cusps. In silver staining, erythrocytes will appear as small black rings, thrombocytes as more granulated material. Fig. 51 shows one of these plaques, some portions of which have been overgrown by new endothelium even at this stage. The new endothelia are characterized by very faint staining of the cement-lines. Some of them are still dissociated, i. e. they do not form a uniformly coherent layer yet. (Similar developments were also found in other traumatized regions of vascular walls than those of valves *Gottlob* and *Zinner,* 1962.)

2. Thrombosis of valvular sinus. Valve-pocket thrombosis is recognized by the fact that the cusp is largely adherent to the parietal endothelium. In earlier stages also hemoglobine is found under the valvular cusp.

3. Adhesion of valvular cusps. Such adhesions may cover a comparatively broad area. Alternatively, they may be of tipped shape (illustrated in Fig. 52 A).

4. Widening of commissure. Fig. 52 A shows a moderately widened commissure – the acute angle has been blunted –, in which there are also smaller adhesions to the parietal endothelium. Fig. 52 B shows a more massively widened commissure. The valvular cusps no longer touch each other.

5. Shrinking of valvular cusps. Fig. 53 A, B shows a valvular cusp, the end of which passes on into a parietal plaque. The cusp appears narrowed. This process may ultimately result in a variety in which cusps are missing altogether and there are only:

6. Valvular bulges. Such valvular bulges are shown in Fig. 53 A, B, and Fig. 54 A.

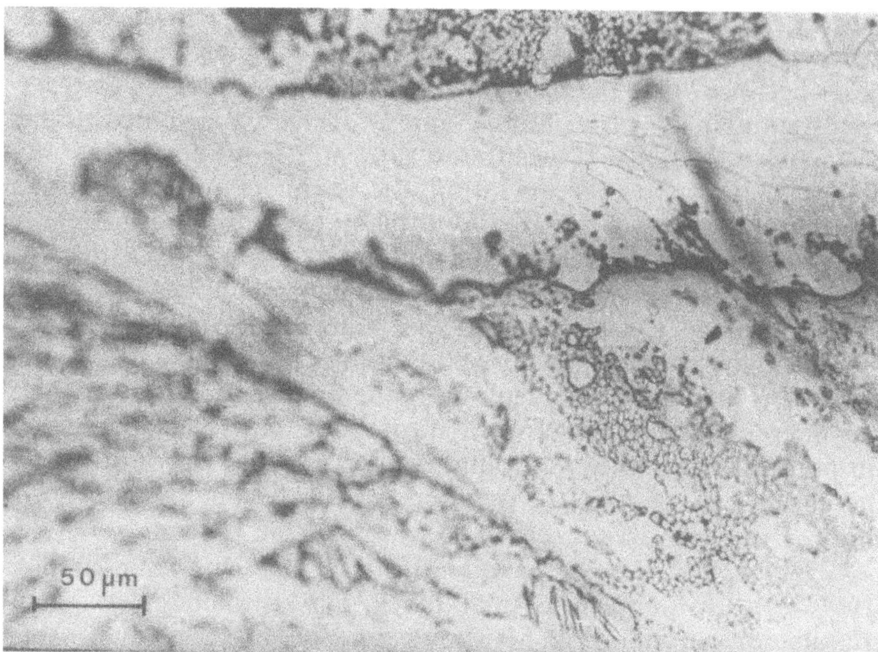

Fig. 51. Valve from jugular vein of rabbit, 24 days after thrombotization. The commissure would be located beyond the left margin of the picture. On the lower valvular cusp there are erythrocyte plaques, some of which are overgrown by new endothelial cells (right lower corner)*

7. Bucket-handle-like bulges. Fig. 56 B shows a number of bulges, some of which are adherent while others are of bucket-handle-like appearance. Such a bulge is shown in Fig. 58 at two different levels.

In Fig. 58 A the bulge is covered by endothelium. In some areas even two layers of endothelium are recognizeable. If the focus of the microscope is lowered, the level of the parietal endothelium (clearly visible in Fig. 58 B) is discernible. We may assume, therefore, that the bulge projects into the lumen of the vessel in a bucket-handle-like manner.

8. Polyps. Fig. 55 A and B show – on two different levels – a polyp found near a valvular commissure. Fig. 55 A shows the parietal endothelium and Fig. 55 B shows the endothelium on the cupula of the polyp, which is largely covered by endothelium already. Another

---

* Figs. 51 to 55: Asterisked pictures reproduced from: Gottlob, R., Kimmel, A.: Post-thrombotic changes of venous valves. Virchows Arch. Path. Anat. *358,* 249 (1973)

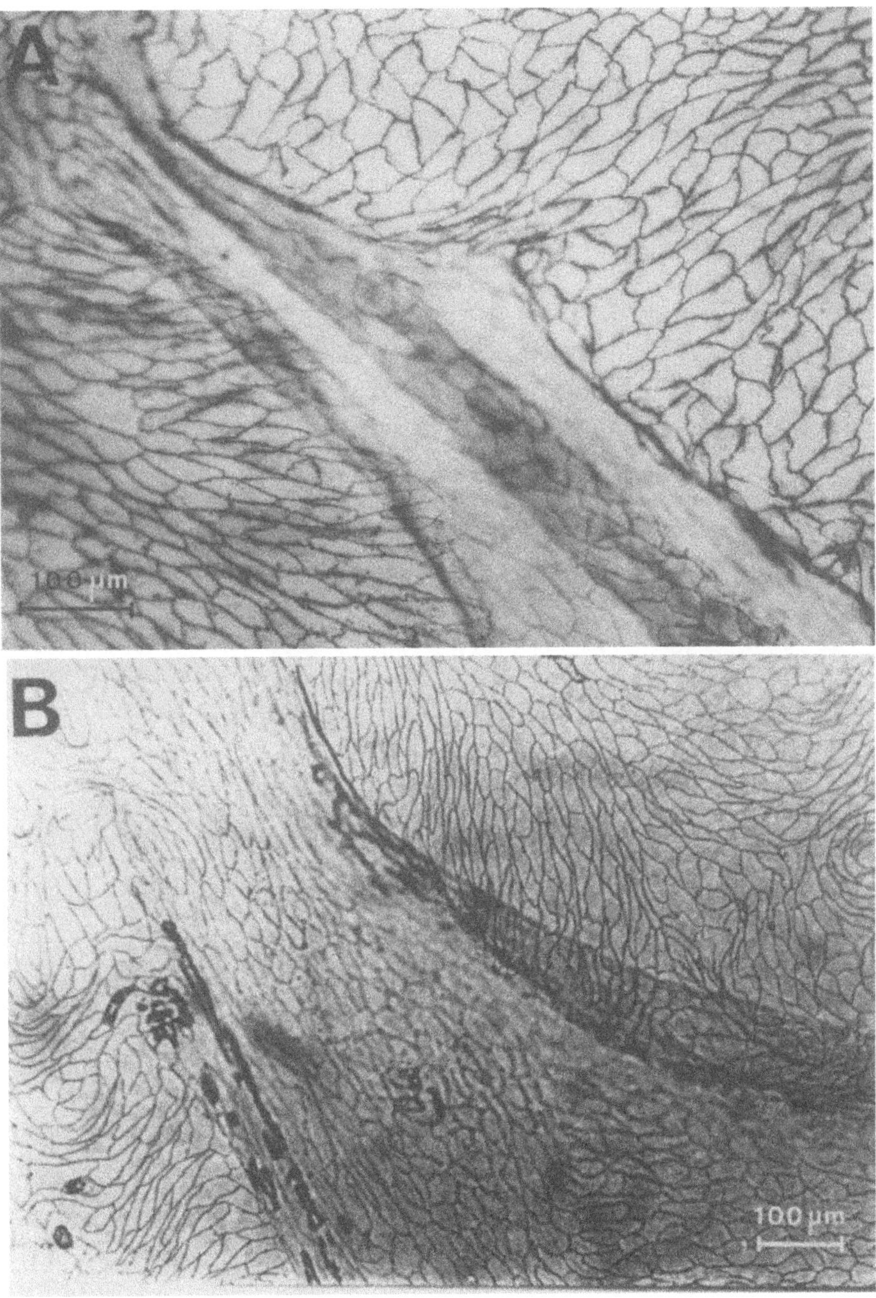

Fig. 52. **A** 11 days after thrombotization. Moderately widened commissure, partly adherent to the venous wall.* **B** 111 days after thrombotization. Massively widened commissure causing incompetence of venous valve*

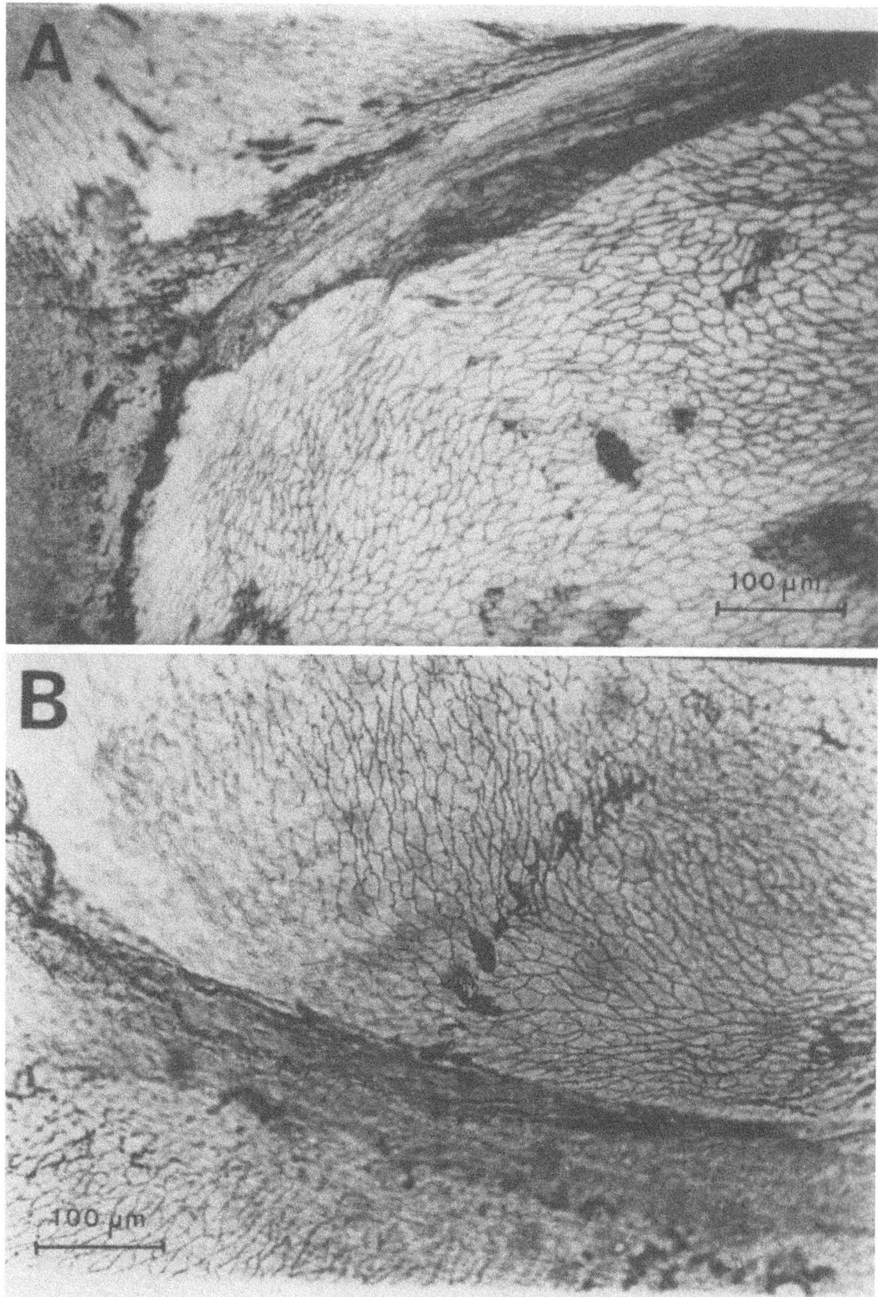

Fig. 53. **A** 7 days after thrombotization. A shrunk cusp passing on into a parietal thrombus (left).* **B** 111 days after thrombotization. The valve has been reduced to a bulge covered by endothelium*

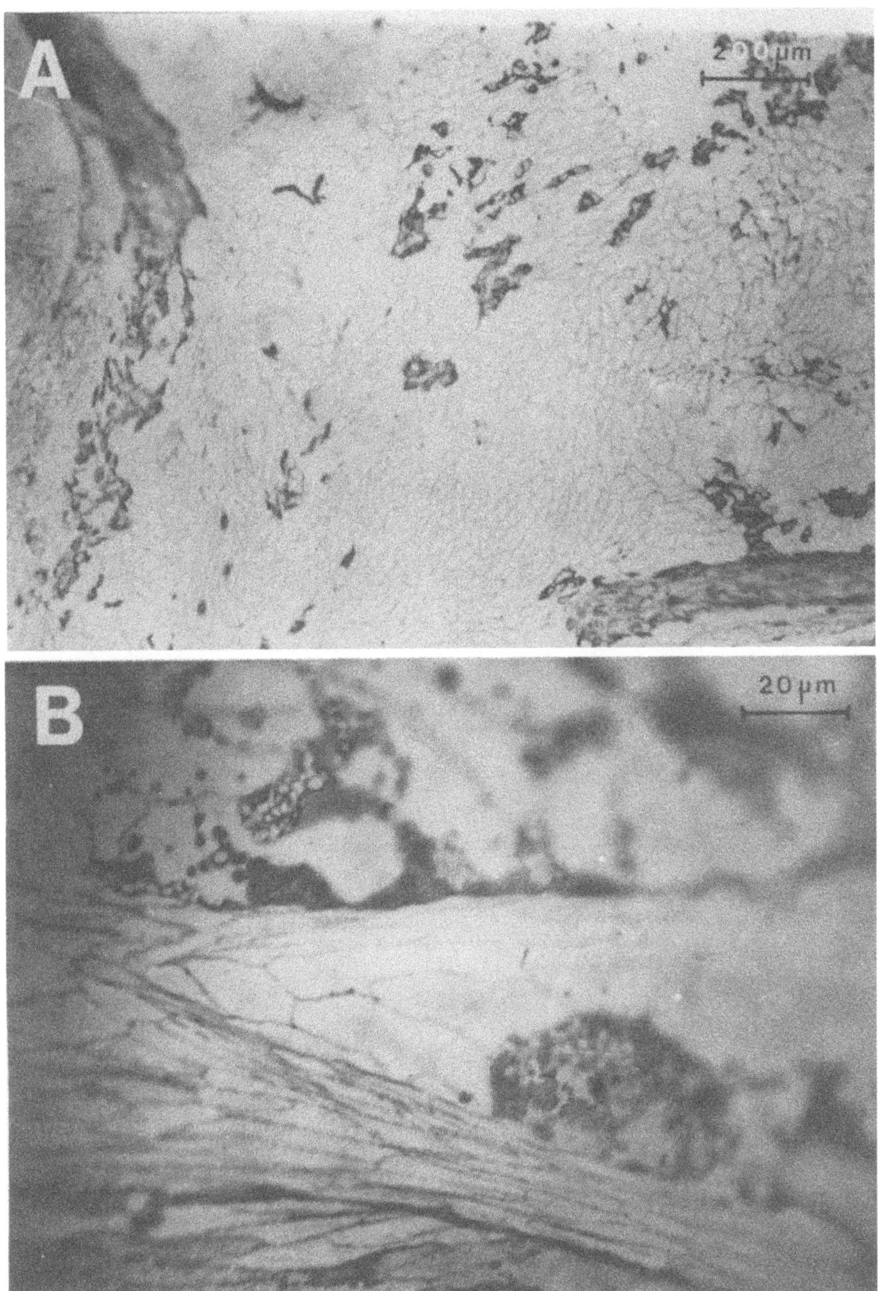

Fig. 54. **A** Two bulge-type valvular rudiments, not touching each other. 111 days after induction of thrombosis. **B** Small ball-shaped thrombus near commissure. 9 days after thrombotization

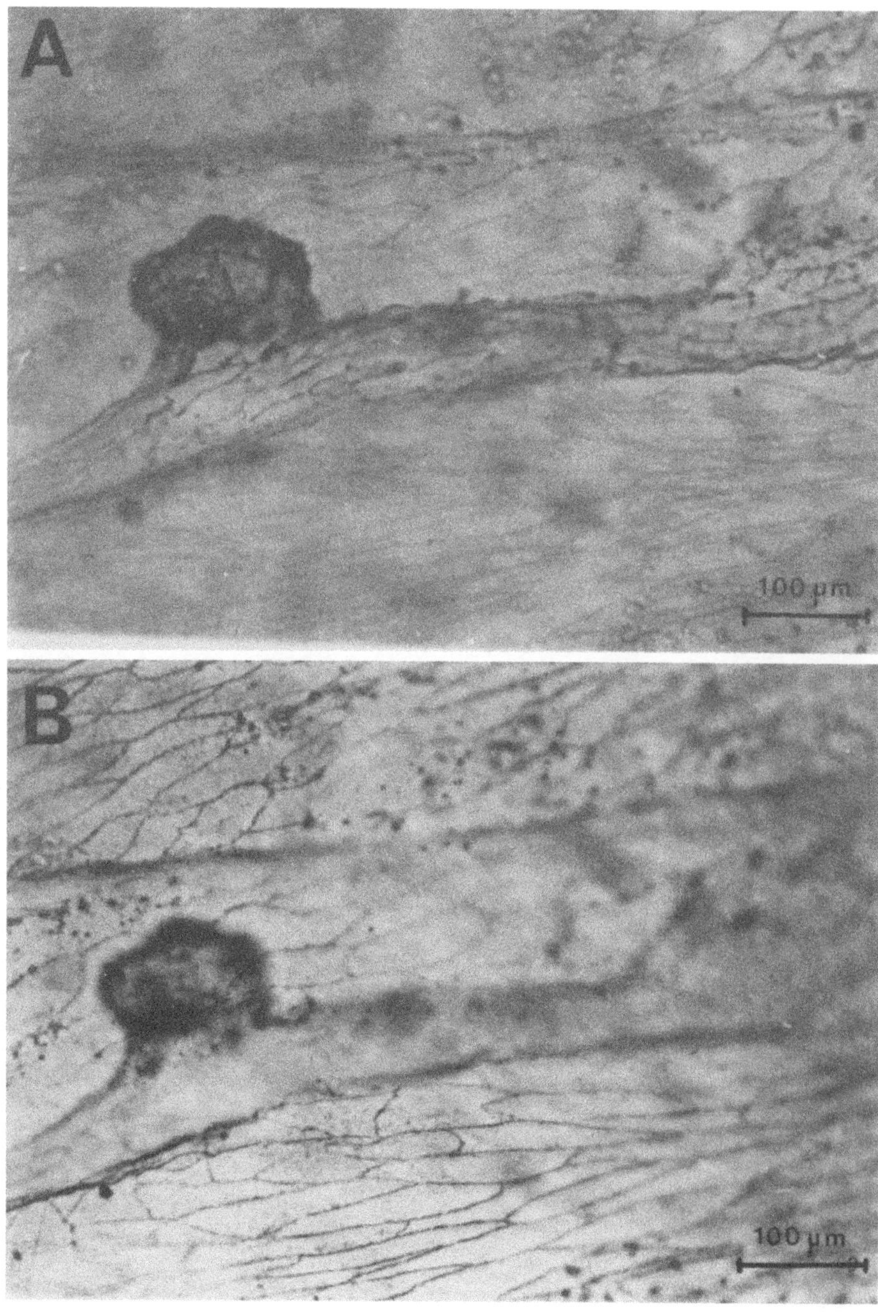

Fig. 55. Small polyp on shrunk valvular cusp. **B** On level of parietal endothelium.* **A** When the focus of the microscope is raised, the endothelium of the valvular cusp as well as the endothelial cover of the polyp is visible. 111 days after thrombotization*

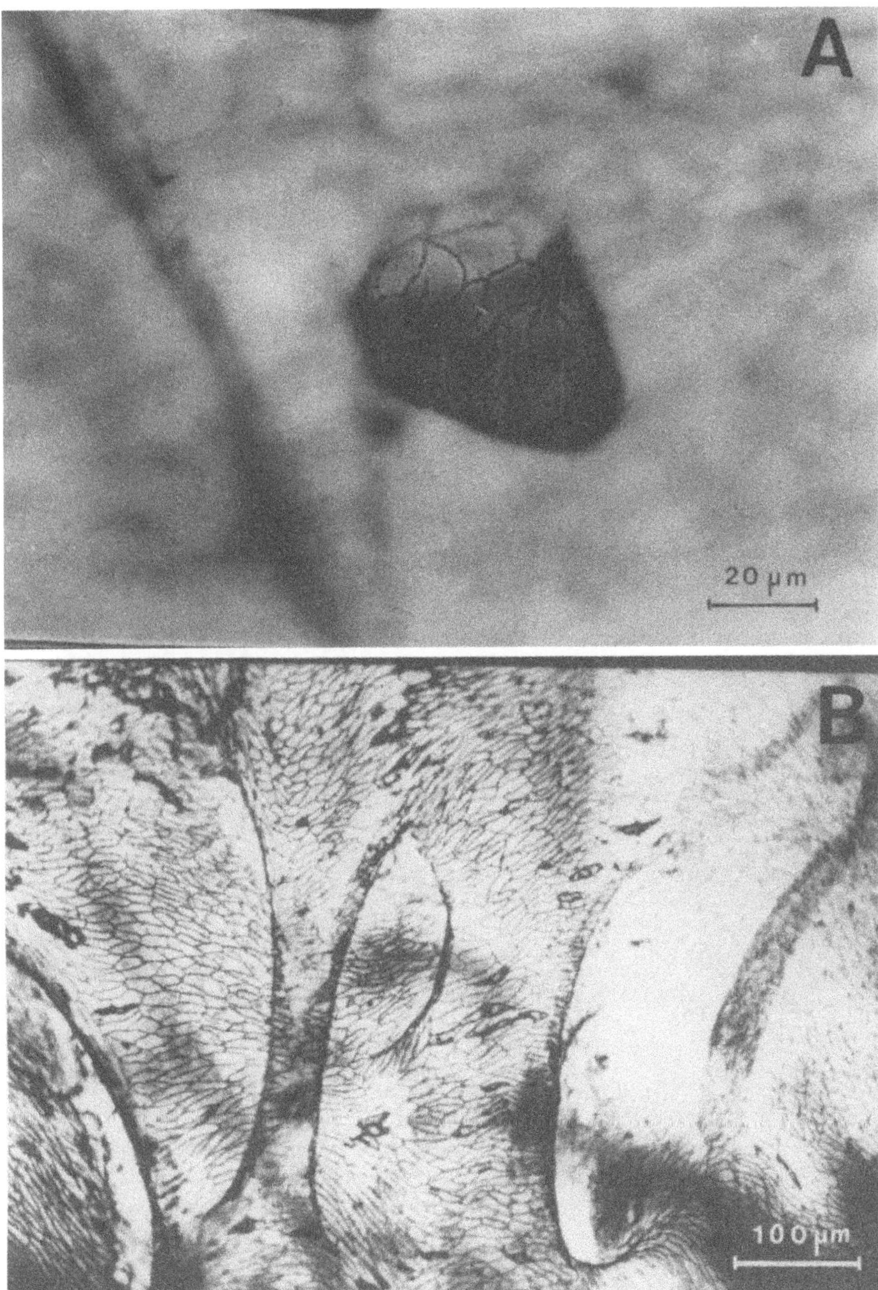

Fig. 56. **A** Polyp near valve, observed 3 months after mechanical trauma.
B Bucket-handle-like bulges. Probably organized thrombi

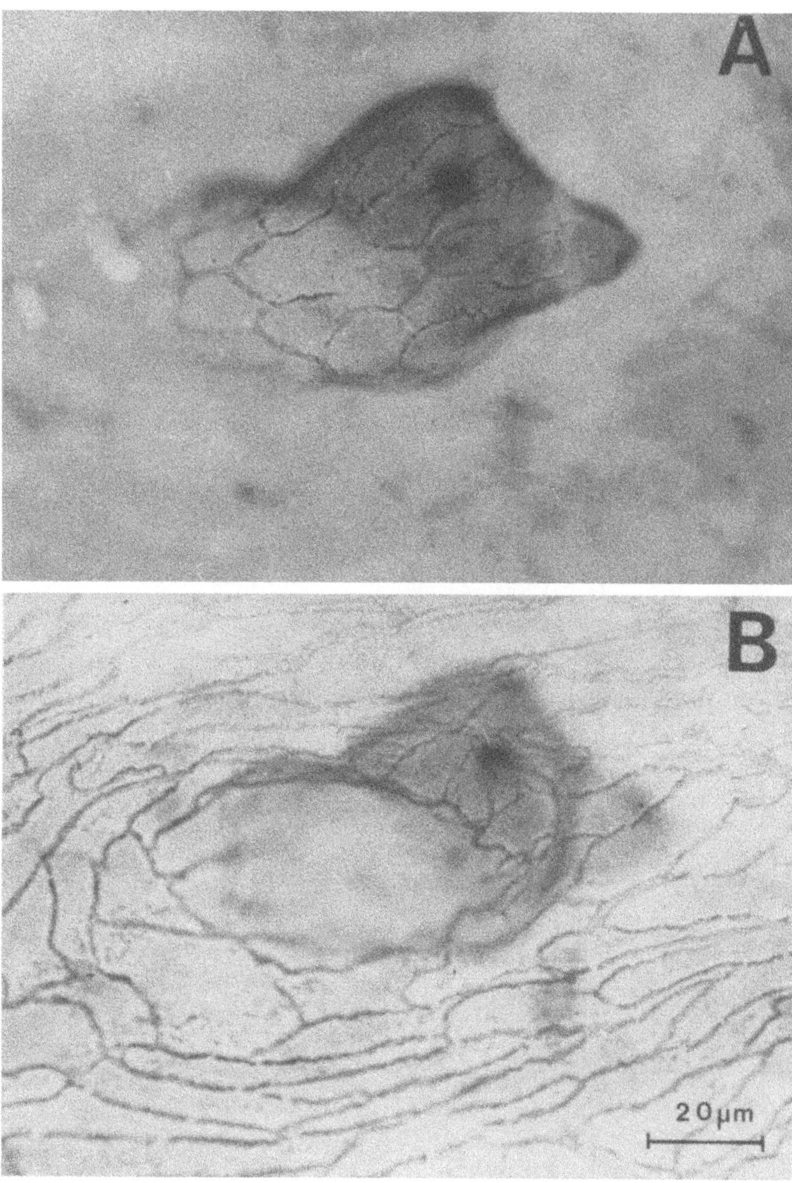

Fig. 57. Polyp 3 months after mechanical lesion near valve. **B** Endothelial layer of vascular wall. The neck of the polyp corresponds to a zone free from endothelium. **A** Same area under raised focus of microscope. The endothelial layer covering the polyp is visible

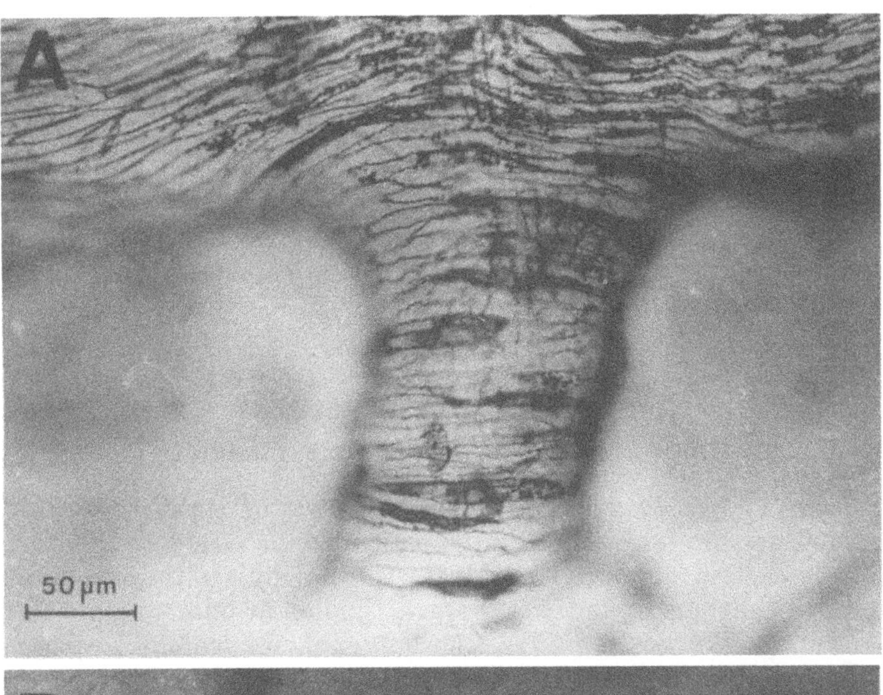

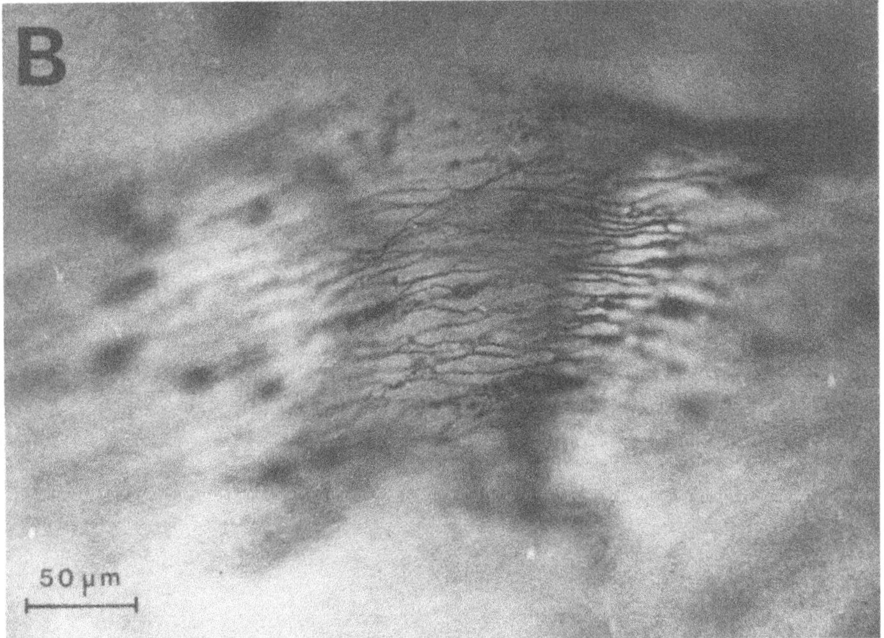

Fig. 58. A bucket-handle-type bulge has occurred after mechanical trauma in valvular region. High power magnification. **B** Endothelium of vascular wall = deep layer. **A** Same area under raised focus of microscope. Level of the surface of the polyp. The superficial endothelium of the polyp is clearly visible (its course is horizontal in Fig. A). The endothelium of the lower side of the polyp is somewhat less clearly visible (its course is vertical in Fig. A)

polyp covered by endothelium is shown in Figs. 56 A and 57, whereas Fig. 54 obviously illustrates an early stage of a polyp, during which it is not yet overgrown by endothelium. One discerns deposited elements of blood and may assume that such polyps originate from small thrombi, which are covered by endothelium later on.

All observed forms of post-thrombotic lesions of venous valves result in valvular incompetence. Forms of clear-cut stenosis were found only in specimens of early thromboses before spontaneous resolution of the thrombotic masses.

H. Valves in Varicose Veins

a) Macroscopic Findings and Conventional Histology

Edwards and *Edwards* (1940) compared valves of 51 phlebologically normal persons with those of 106 patients with varicose saphenous veins. – Unless there was a past history of phlebitis of the saphenous vein, the changes found were largely similar in both primary and secondary varices (Fig. 59). The bulge of the commissure was flattened, and there was evagination of the interspace between the two cusps or even aneurysmatic expansion of this region so that the insertions of the cusps were markedly separated. The elastic membrane between the two commissures was thinned and interspersed by numerous fenestrae.

These changes are followed by further ones, which are to be interpreted in terms of a reparative process. There will be ingrowth of

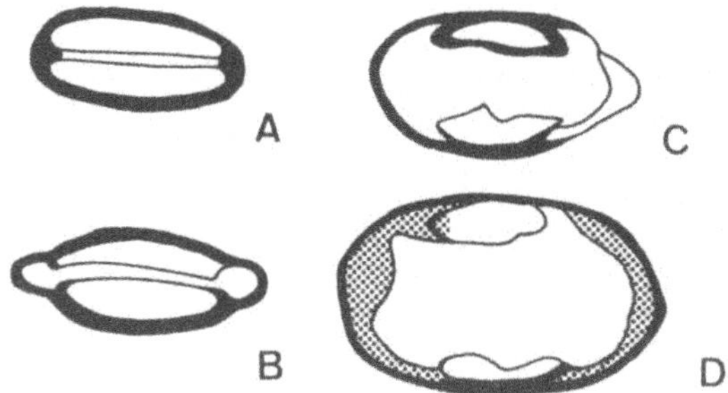

Fig. 59. Progressive development of varicose changes shown in venous cross-sections. Incompetence of valves is to be attributed to venous dilation rather than primary valve lesions. **A** Normal vein. **B** Early stage. The vein is dilated in no other region than in that of the commissure. **C** and **D** The cusps are far apart. **C** The commissure contains reparative tissue. **D** Such tissue is found in both commissures but cannot efficiently compensate for diastasis of the cusps [From Edwards and Edwards: Amer. Heart J. *19*, 338 (1940)]

fibroblasts and ample quantities of smooth muscle cells as well as some capillaries, penetrating into the luminal portions of the dilated and expanded commissure through the fenestrae of the elastic membrane. The fibroblasts may form masses of collagen, fresh elastic tissue may develop. The newly formed tissue may proceed across the contact area of each of the cusps or even invade the subvalvular region. The expansion of the region of the commissure is partly compensated by the invasion of new tissue. This new tissue may form a thin layer under the endothelium, or heavy cushions filling the evaginated region of the commissure and occasionally arching forward into the lumen. Adhesions with valvular cusps develop relatively late. Through such adhesions the valves may be drawn toward the surface of the cushions or directly to the evaginated venous wall. Still, the cushions are unable to narrow the lumen to such an extent that the effect of evagination – i.e. valvular incompetence – is compensated.

One of the consequences of these changes is that the course of the valvular cusps changes. They may be kinked or even turned round so as to confront the direction of the bloodflow.

Leu (1971) describes a valvular cusp in a varicose vein that was considerably thickened and shrunk.

Compared to post-thrombotic changes, the changes of valvular cusps observed in varices were of relatively minor nature. For this reason, *Homans* (1916), *Bernstein* (1927), *Nicholson* (1923) and *Leu et al.* (1979) are of the opinion (also held by *Edwards* and *Edwards* [1940]) that the changes of valves observed in varicose veins are not the cause of the disease but merely the sequelae of primary venous dilation. A number of weighty arguments may be brought forward to substantiate the assumption that the primary cause of varicose veins is to *be seen in dilation of the venous wall.* Firstly, varices occur during the first months of pregnancy, during which there is not yet venous stasis caused by the enlargement of the uterus, and secondly, the observation that the arm veins of patients with varicose veins are more dilatable than those of phlebologically healthy persons (*Szotér* and *Cronin,* 1966). On the other hand, one should not disregard the findings of *Ludbrook* and *Beale* (1962), who observed that in obviously phlebologically healthy subjects whose ancestors had suffered from varicose veins an absence of valves above the sapheno-femoral junction occurred significantly more frequently than in subjects without such hereditary taint. These findings established by invasive pressure measurement were confirmed by *Reagan* and *Folse* (1971), who used a method of percutaneous flow measurement. They described "descending sequential valvular incompetency". Finally it must be stressed that valves in varicose veins usually are obtained exclusively from extirpated peripheral veins. While in those veins changes of the cusps may indeed be secondary, this might be due to valvular loss or missing valvular anlage at more proximal sites, i.e. above the sapheno-femoral junction. Still, as a rule, these vessels are evasive of direct observation.

On the basis of information available we would not rule out any of the following three potential causes of the development of primary varicose veins:

1. weakness of venous wall;

2. loss, or at least incompetence, of valves at more proximal sites; and

3. a combination of (1) and (2) above.

b) Own Findings in Subterminal Valves of Long Saphenous Veins (Silver-stained En-face Preparations)

During high saphenous ligature performed on patients suffering from primary varicosis, the uppermost portion of the long saphenous vein was resected, avoiding any mechanic traumatization. The vessel was cut open, spread and silver-stained (in the manner described on pp. 42 ff.) during the operation. The specimens obtained (*Gottlob et al.,* 1975) showed a whole variety of valvular changes similar to those found among post-thrombotic valves. In no more than 5 out of 21 cases was one cusp of the long saphenous vein found to be largely normal. Nevertheless, the method used does not allow safe conclusions concerning the respective partner-cusps. Besides, incompetence due to dilation of the vessel cannot be outruled. The following pathologic changes were observed: shrinking of cusps; broad-based adhesions between valvular cusp and vascular wall; narrow areas of synechia between cusps and wall of the vessel; widened commissures; bulges and cascades in the valvular region and irregularities of the endothelium in the valvular region. Defined dilation in the examined portion of the long saphenous vein was found in 11 out of 21 cases; in 5 of the 11 cases there was genuine aneurysm. The distribution of findings is shown in Table 2 (p. 120).

In one other case, which is not included in Table 2, bulges protruding into the lumen in a bucket-handle-like manner and a polyp covered by endothelium were found (Figs. 65 and 68, *Gottlob* and *Saghir,* 1976).

Due to the fact that prior to resection of the saphenous segment for our studies, the main trunk was ligated closely to the junction, it is to be assumed that the uppermost (terminal) saphenous valve, which is

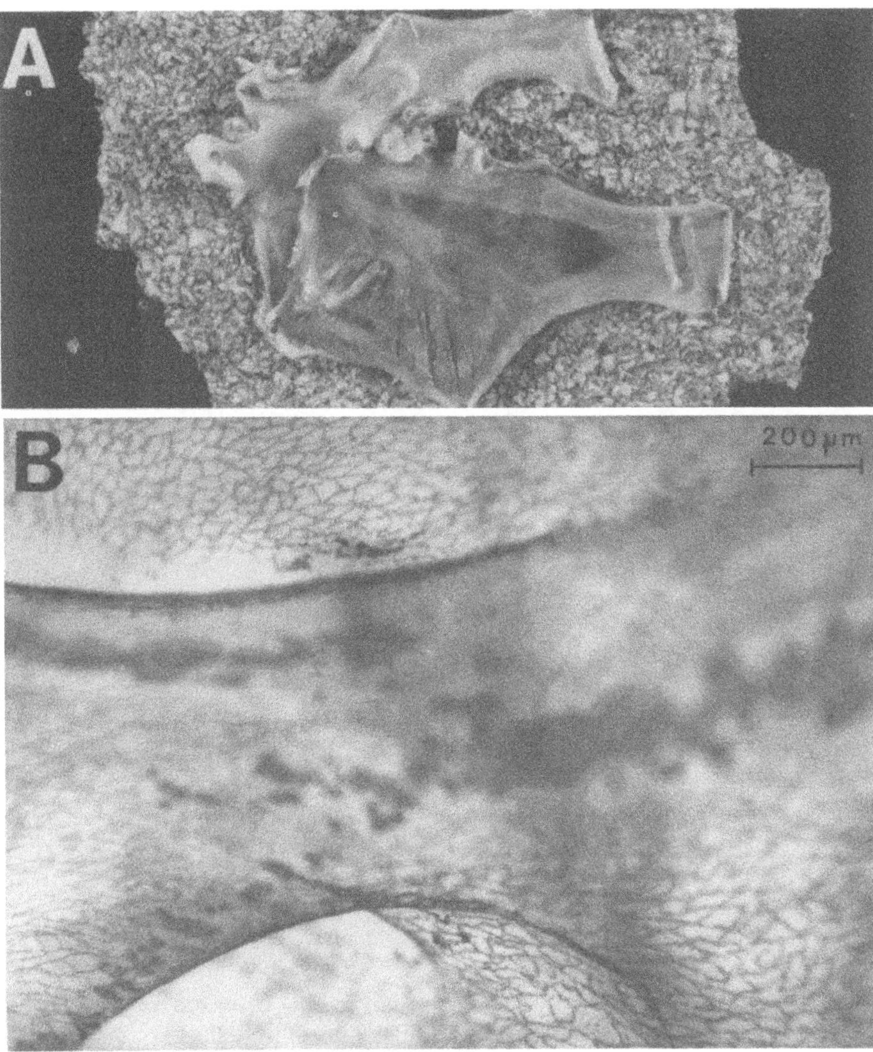

Fig. 60. **A** Aneurisma-like dilation of internal saphenous vein in immediate vincinity of junction. Undilated duplicature of vein (top). Note irregularities of intima in aneurismatic area. **B** Moderately widened commissure. [Fig. A from Gottlob, R., Donas, P., El Nashef, B.: Zbl. Chir. *100*, 1306 (1975)

located at this particular site, was seized by the ligature so that the valve examined was likely to be the *subterminal* valve of the internal saphenous vein, unless (as shown in Fig. 9 B) the terminal valve was located at an extraordinarily low site.

The specimens prepared by the en-face method indicate that valvular lesions seem to occur quite frequently in superficial venous

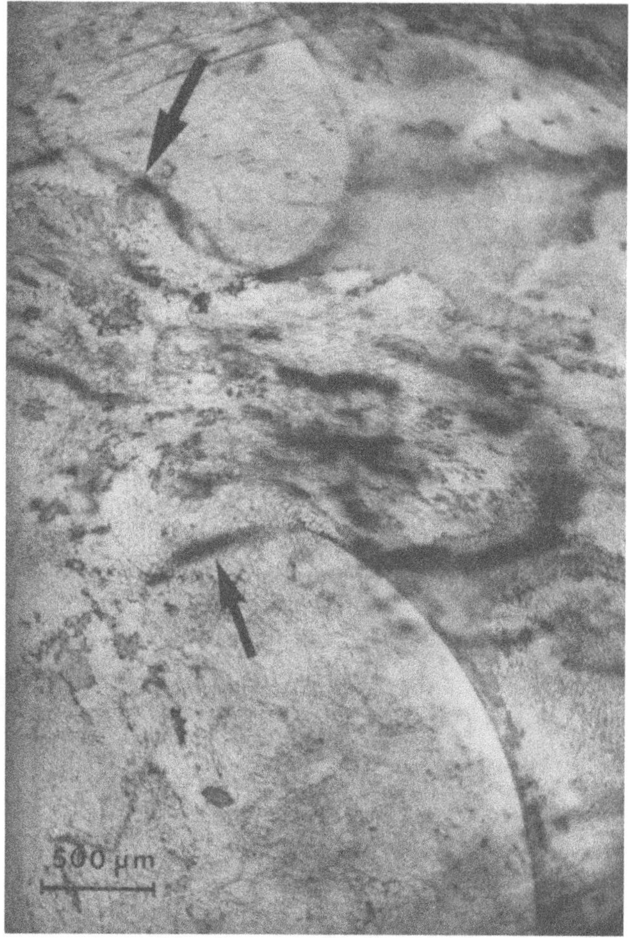

Fig. 61. Lengthy adhesion (arrows) of cusp (right) to vascular wall. [From Gottlob, R., Donas, P., El Nashef, B.: Zbl. Chir. *100* 1314 (1975)]

trunks even under conditions of primary varicosis. Unfortunately, our findings are not conclusive with respect to the question as to which extent the development of varicose veins is to be attributed primarily to valvular lesions or whether the primary cause is venous dilation, which in turn might have been caused by increased intraluminal pressure or by degeneration of the vascular wall.

In the valves of varicose saphenous veins we found the same changes that were observed after spontaneous lysis of experimental thrombosis. This leads to the conclusion that changes of valves in varicose veins are caused by the organization of local thrombi and microthrombi.

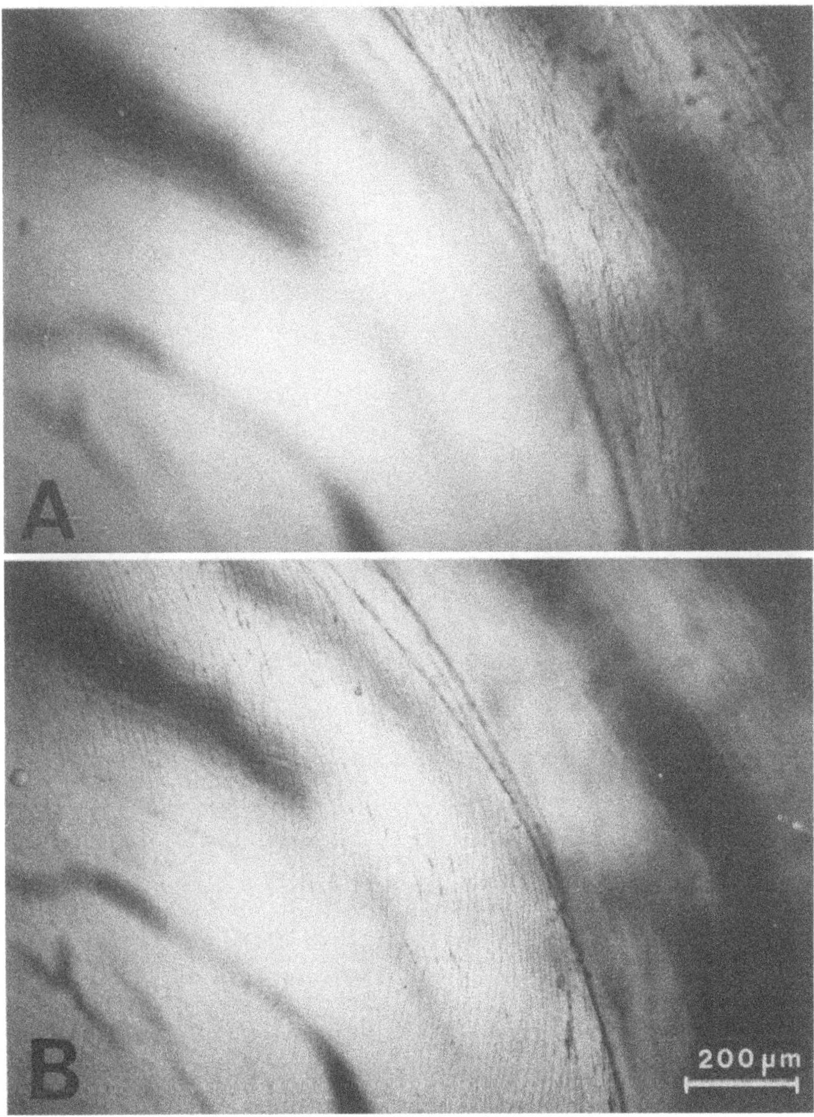

Fig. 62. Cusp completely "soldered" to vascular wall. **B** On level of vascular wall. **A** On level of valvular cusp. The cusp forms a step to the vascular wall. Endothelial layers not present under surface of cusp. [From Gottlob, R., Donas, P., El Nashef, B.: Zbl. Chir. *100*, 1313 (1973)]

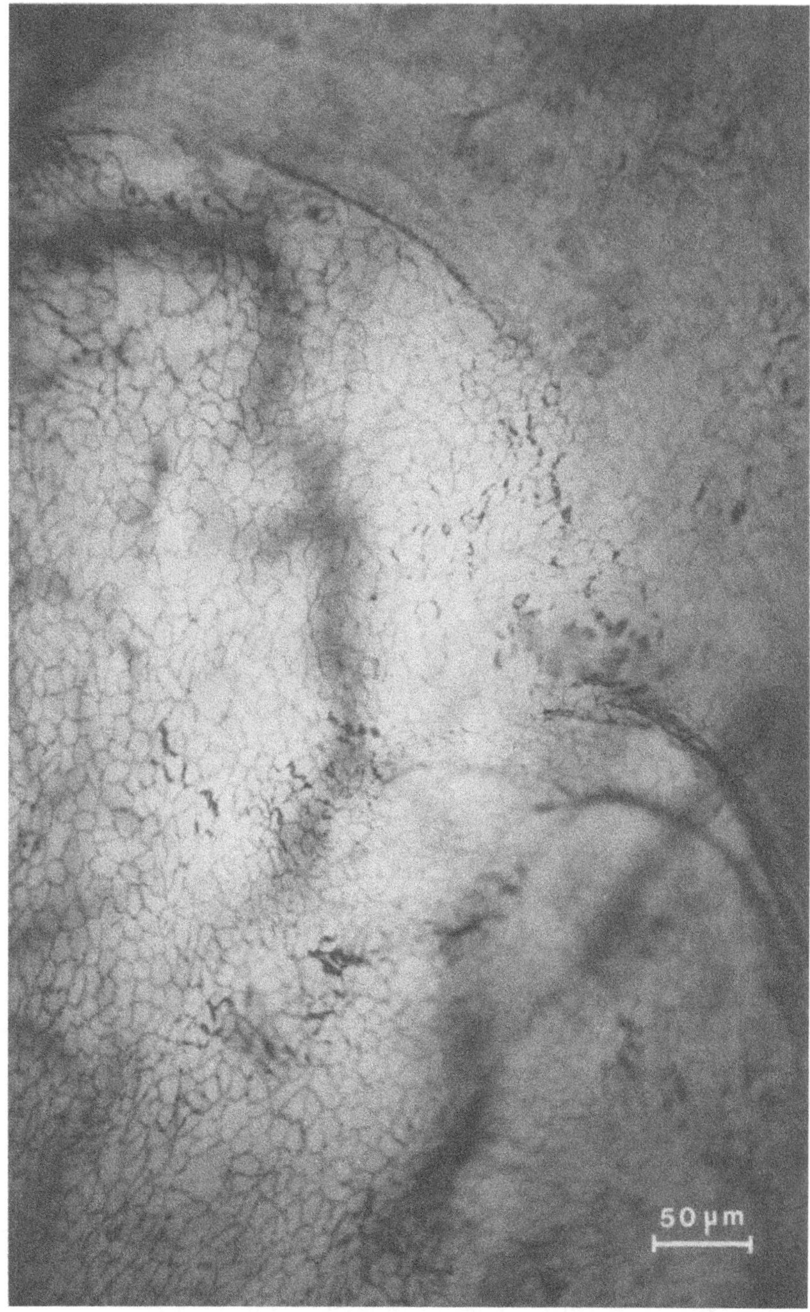

Fig. 63. Valvular cusp reduced to irregularities of endothelium. Commissure extensively enlarged

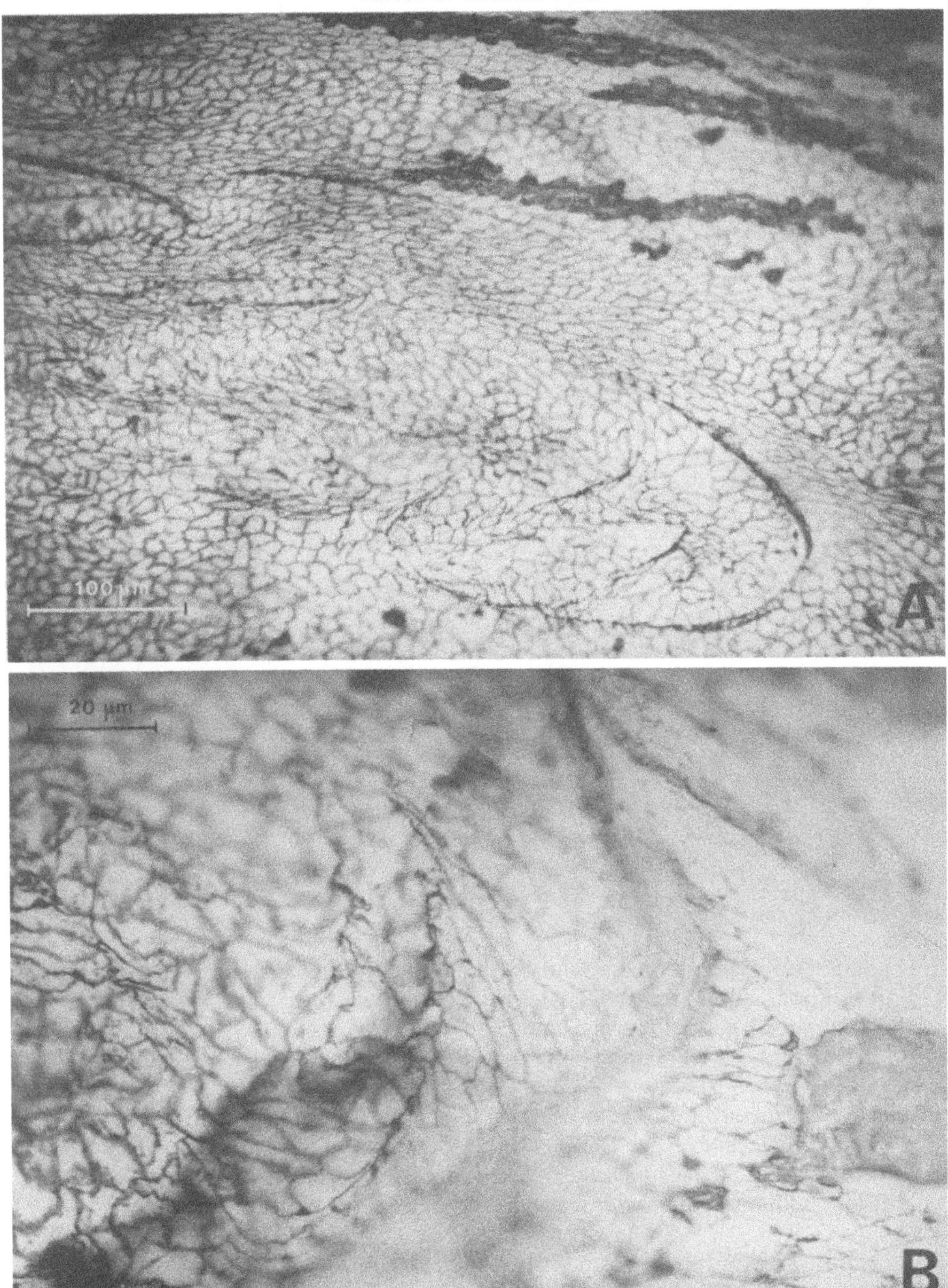

Fig. 64. Irregular patterns of endothelium. **A** In the region of a aneurism-like dilation near sapheno-femoral junction. **B** In the region of a valvular sinus, whose cusp has completely shrunk

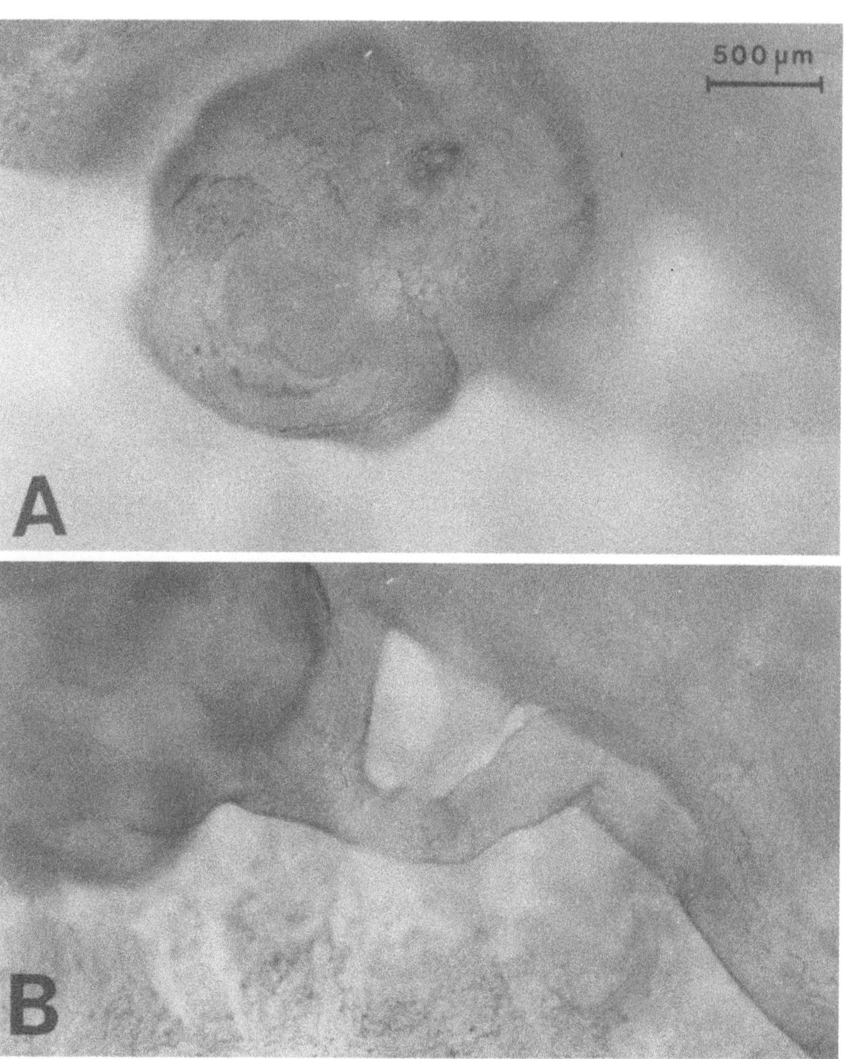

Fig. 65. A polyp and bucket-handle-like bulges at the edge of a valve. Cusp in the upper right corner. **B** At level of endothelium of vascular wall. The polyp in the upper left corner is only shown in part. **A** Polyp visualized in the centre of the picture on the level of the endothelium covering the polyp

---

Fig. 66. Bulges running toward the polyp from the edge of valvular cusp, shown at different levels. **C** Level of endothelium of vascular wall. **B** Level of endothelium covering the lower side of left broad bulge. **A** Level of endothelium covering upper side of left bulge. [From Gottlob, R., Saghir, F.: Phlebol. u. Proctol. *5*, 1 (1976)]

Fig. 67. Tricuspid valve from female patient with varicose veins, saphenofemoral junction. **A** Macroscopic image of valve; the site of the former valvular pockets is clearly recognizable; it is indicated in scheme **B.-D, E** and **F** Microphotograms of valvular pockets. In **D** merely the site of the former insertion of the cusp is recognizable (arrows). In **E** a small valvular rudiment is visualized, and in **C** and **F** the shrunk cusp of the third valve is shown on two different levels

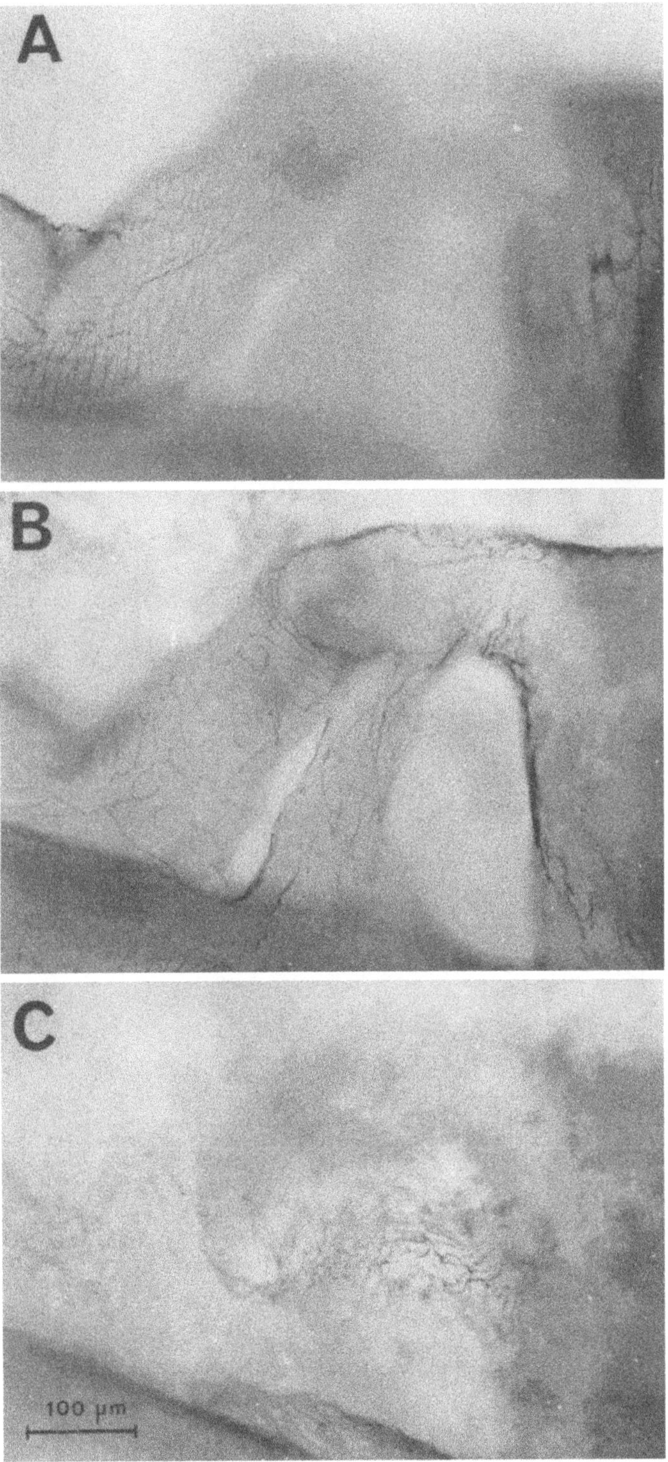

Fig. 66 A–C

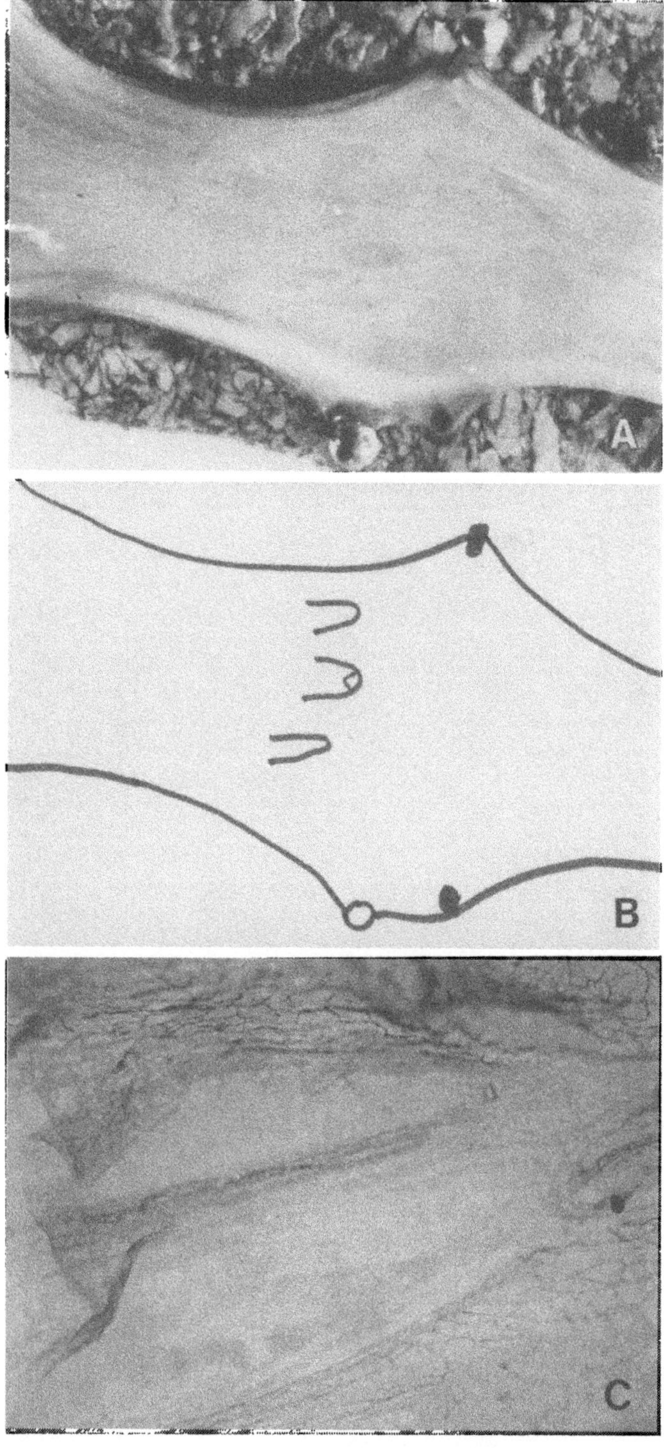

Fig. 67 A–C. (Legend see p. 116)

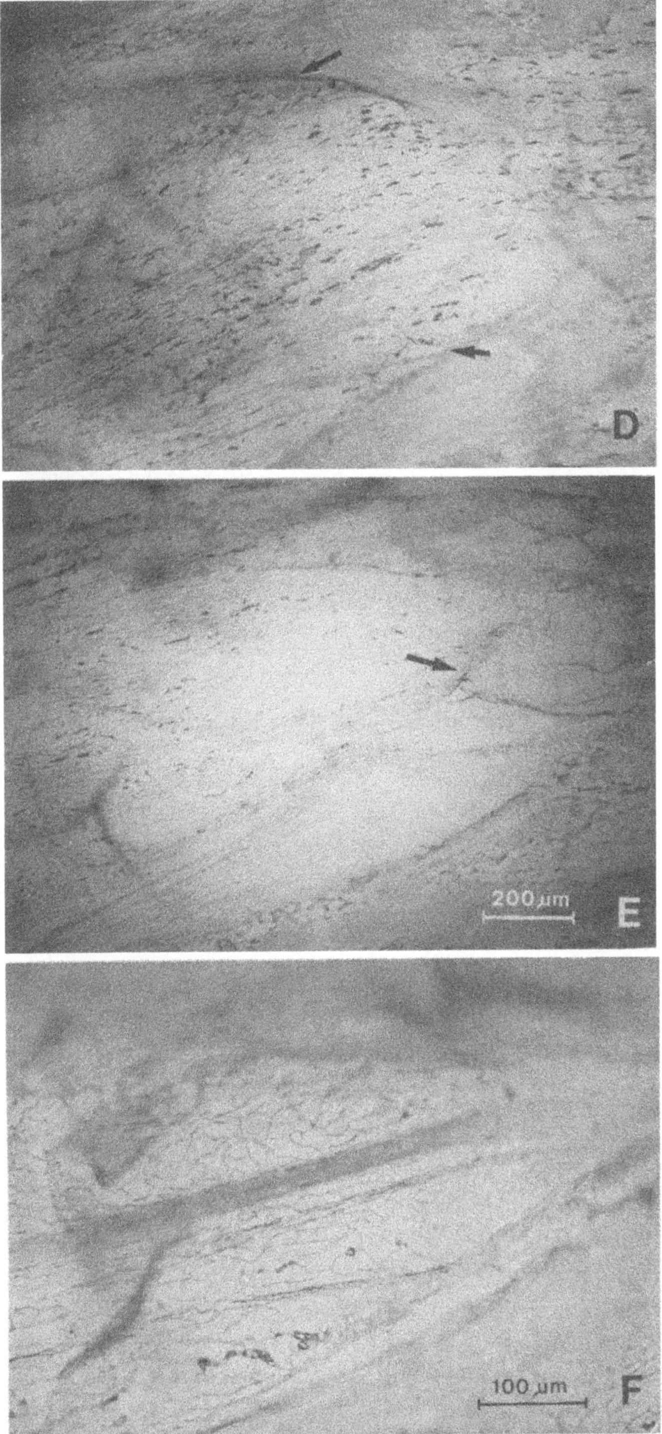

Fig. 67 D–F. (Legend see p. 116)

Table 2. *Findings comprising 22 terminal or subterminal valves of long saphenous veins from patients suffering from primary varicosis*

No.	Largely normal cusp	Cusp shrunk	Cusp adherent over lengthy basis	Short synechia	Widened commissure	Bulges, cascades	Irregularities of endothelium	Dilatation (+), Aneurism (++)
1	+							++
2	+						+	+
3			+			+		+
4		+			+			
5	+ (In duplicature, absent in main trunk)							
6	+						+	
7			+			+	+	
8			+			+		
9		+			+			+
10		+			+	+		++
11	+							+
12			+			+	+	++
13	+							
14				+	+			+
15	+ (Plaque on commissure)							
16		+					+	
17		+						
18			+			+	+	
19				+		+		++
20	No valve found							
21				+	+		+	+
22	Bucket-handle-like bulges and broad polyp (Fig. 65, 66)							

One of the changes observed most frequently after spontaneous lysis of experimental thrombi, was widening of the commissure (Figs. 60 and 63) causing valvular incompetence.

Edwards and *Edwards* (1940) observed that massive dilations of the venous wall between the insertions of the cusps occurred in varicose veins. We found these findings confirmed, e.g. in the varicose tricuspid valve shown in Fig. 67. The area of the commissure is definitely the weakest region of the venous wall. In addition to the tendency to dilation, there seems to be an enhanced tendency towards adhesion between valvular cusp and venous wall in this region. This is perhaps due to the fact that the excursions of the valvular cusp are minimal in the region of the commissure. Thus, widening of the commissure may occur even without dilation, as we have found after spontaneous lysis of experimental thrombi.

2

Pathologic Function

Destruction or incompetence of individual venous valves need not necessarily have clinical consequences. On the other hand, it may result in a "chain reaction" by causing an increase of pressure in the distal portion of the vein, which may lead to dilation and, thereby, incompetence of the valve located at the next site upstream (see p. 23). The specific clinical significance of individual venous valves varies. We may assume that the destruction of an individual valve – e.g. by thrombosis of its pocket – need not have clinical consequences if there are still healthy valves proximal to it. A key position is held by those valves whose location is centralmost since they are not protected by other valves at sites proximal to theirs. The valves separating the deep leg veins from the superficial leg veins are also in a key position as severe differences of pressure occur at their sites. In contrast to the superficial veins, the deep veins are protected by surrounding striated muscles so that they are only rarely subjected to unphysiologic dilation. A key position of particular significance is held by the valves at the entrances of the long and short saphenous veins into the deep leg veins. The valves located at these sites are obviously subjected to particular strain, especially in the absence of more proximal competent valves. The entrances of the long and short saphenous veins may also be regarded the largest perforating veins.

Diseases occurring in combination with incompetence of venous valves are: varicocele, post-thrombotic syndrome, and primary varicosis.

A. Varicoceles

Although it is the authors' intention to deal mainly with veins draining the blood from the legs in this book, we must dedicate the following chapter to the valves of the internal spermatic vein (testicular vein) in view of their enormous clinical significance.

Valves were found in the gonadic veins (internal spermatic vein, ovarian vein) by *Rivington* as early as 1873. In later years, examinations of cadavers were carried out by *Fagarasanu* (1938) and *Ahberg et al.* (1965, 1966, 1966a). *Fagarasanu* observed that in 30 cadavers examined valves were absent in the gonadic veins in more than 50% of the cases. According to *Ahberg et al.*, valves were found in 135 out of 185 specimens. Gonadic veins of male individuals were significantly more frequently valveless than those of female individuals, and valves were absent more frequently on the left side than on the right side. Out of 135 specimens with recognizable valves, 122 showed a valve directly at the entrance of the gonadic vein into the left renal vein or into the caval vein. The competence of valves was evaluable in 108 specimens. The results of evaluation are shown in Table 3.

Table 3. *Cases of valvular incompetence among 108 evaluated venous valves of gonadic veins*

Sex	Number	Competent	Incompetent
Male	33	26	7
Female	75	40	35
Total	108	66	42

(According to *Ahberg et al.*, 1965.)

So we can state that as far as valves were observed in gonadic veins at all, they were more frequently incompetent in female subjects than in male subjects. Moreover, the diameter of the gonadic veins was larger in female subjects, especially in those who had borne several times.

Varicoceles may be regarded sequelae of valvular incompetence of the internal spermatic vein. In a similar manner as observed with varicosis of the saphenous vein, blood may enter the testicular veins in retrograde direction. From there it passes on into branches of the inferior caval vein via the anterior and posterior scrotal veins and the veins of the deferent duct. Under conditions of varicocele, therefore, there is a private circulation similar to that observed by *Trendelenburg* (1890) under conditions of primary varicosis of the long saphenous vein.

According to *Brodny et al.* (1955), more or less marked varicoceles are found in 10% of men. In 90% of all cases varicocele is restricted to the left side. 8% are bilateral cases, and no more than 2% confined to the right side.

A number of more recent findings have somewhat complicated the assessment of the role of venous valves in varicoceles. By means of angiographic studies of 149 varicoceles *Marsman* (1985) found 27 cases in which the valves of the internal spermatic vein were competent. In these cases the main vessel was bypassed by collateral veins with valvular incompetence. *Mali et al.* (1985) investigated the gradient of pressure between the left renal vein and the caval vein. Cases with low pressure gradients did not show retrograde bloodflow even if valves were absent in the spermatic vein, and the development of varicoceles was neglectable. In contrast, if the gradient of pressure was high, there was considerable retrograde bloodflow in the left spermatic vein, the vein was dilated and there was marked development of varicoceles. There were numerous intermediate stages between the two extremes.

The clinical significance of varicoceles is brought about by a side-effect often encountered in combination with them: *male infertility.*

A connection between varicocele and infertility was first observed by *Russell* (1954). In 1955, *Rob* recommended the therapy of ligating the vein some 5 cm above the inguinal canal. In the same year *Tulloch* observed that the testicles of men suffering from varicoceles were softened, often even on the counterlateral side. In 1965 *McLeod* reported on changes of the sperm of varicocele patients, and *Dubin* and *Hotchkiss* found evidence of histologic changes by testicular biopsies (1969).

What causes infertility of varicocele patients? Physiologic spermatogenesis is possible only at a testicular temperature lower than the basic temperature of the body. This is why men with undescended testicles are infertile. Thermographic investigations (*Comhaire et al.,* 1976; *Gasser et al.,* 1973; *Kormano et al.,* 1970, 1973) confirmed that raised temperatures were prevalent under conditions of varicocele. We may then assume, that infertility is caused by the supply of body-warm blood via the venous system. The question of why there is infertility even in unilateral varicocele is not quite clarified (*Tulloch,* 1955), although it is remarkable in this respect that in 10 cases *Ahberg et al.* (1966a) succeeded in filling scrotal veins of the right side after injection of contrast medium into the left internal spermatic vein.

The method best suited for *diagnosing* varicoceles, in particular in preliminary stages which cannot be proved by optical means or palpation, is angiography. Testicular varices may be punctured or cannulated in order to inject contrast medium. However, with a view to revealing the pathogenesis, the injection of contrast medium into the left renal vein via a catheter introduced into the femoral vein is more suited. Special catheters for this purpose were developed by *Riedl* (1979). They also allow selective catheterization of the spermatic veins (Figs. 68–71).

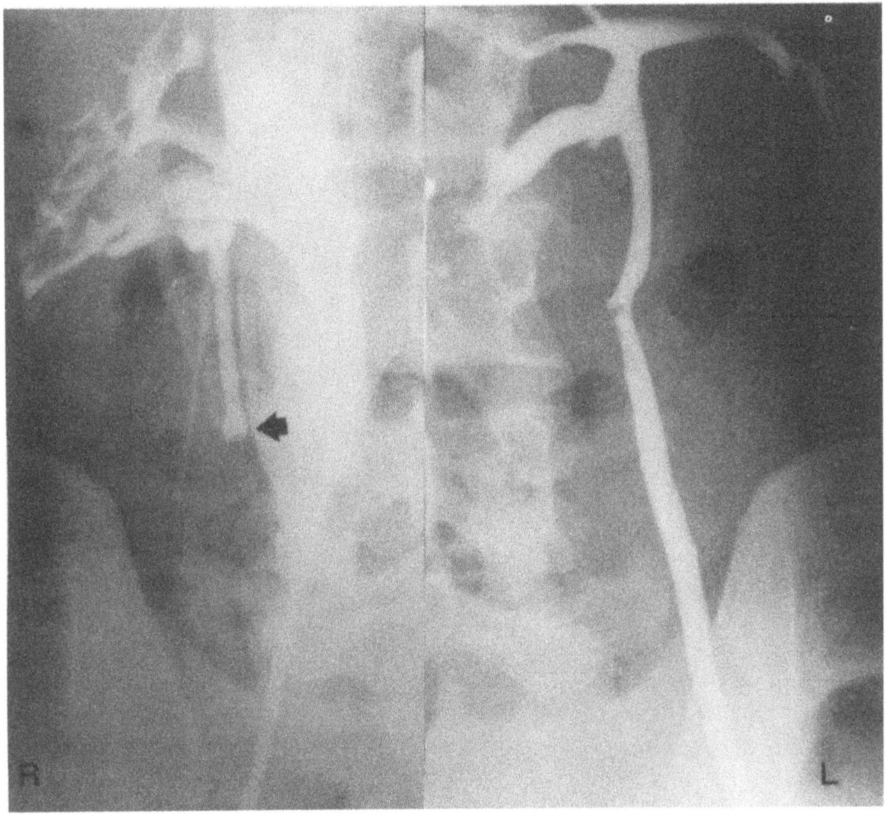

Fig. 68. Competent valve in right internal spermatic vein (↑). Incompetent valve in left vein

Therapeutically, measures of surgery or sclerotization can be resorted to. Besides the operation described by *Robb* (1955), the operation according to *Palomo* (1949), consisting of high ligature of the spermatic artery and vein, is often carried out.

Sclerosing of varicoceles via the scrotum has become obsolete. Better results can be achieved by sclerosing from the spermatic vein, selectively catheterized by the renal vein (*Lima et al.,* 1978). By applying a balloon catheter (*Riedl,* 1979), reflux of the sclerosing solution into the renal vein may be prevented. The sclerosing treatment may thus be carried out while the patient is in a supine position, and the sclerosing solution will not sink into the testicular veins. In this way they will be protected from being sclerosed.

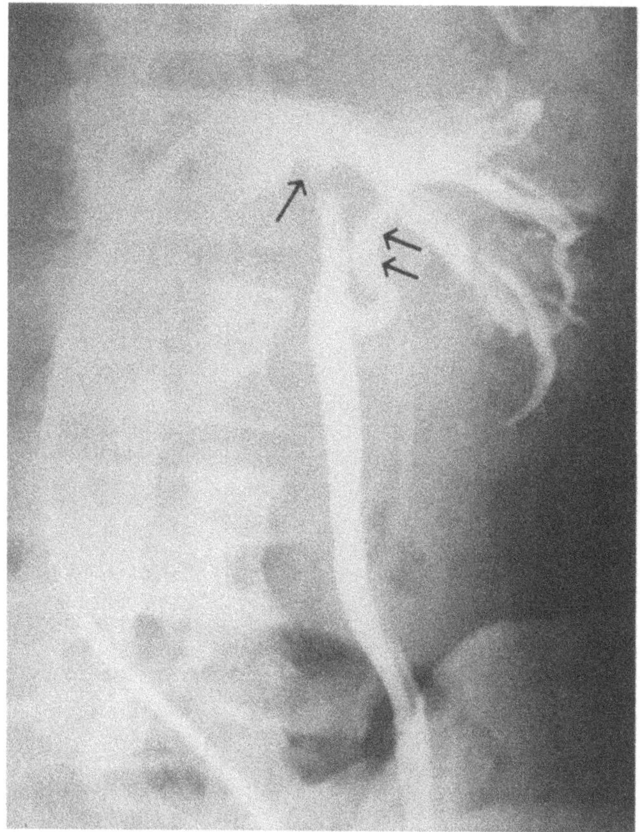

Fig. 69. A competent valve at the cranial end of the internal spermatic vein (↑) is bypassed by an incompetent collateral vessel(↑↑). 3rd degree varicocele

After curing of the varicocele, the difference between testicular and body temperatures increases from levels as low as 0.1° C to largely physiologic differences ranging between 2 and 3° C. The sperm quality of 60 to 80% of operated or sclerotized patients improves, and a gravidity quota of 17 to 40% was observed (*Gasser,* 1971).

The phlebograms of internal spermatic veins were reproduced by courtesy of Dr. P. *Riedl,* Central Institute of Radiology, University of Vienna.

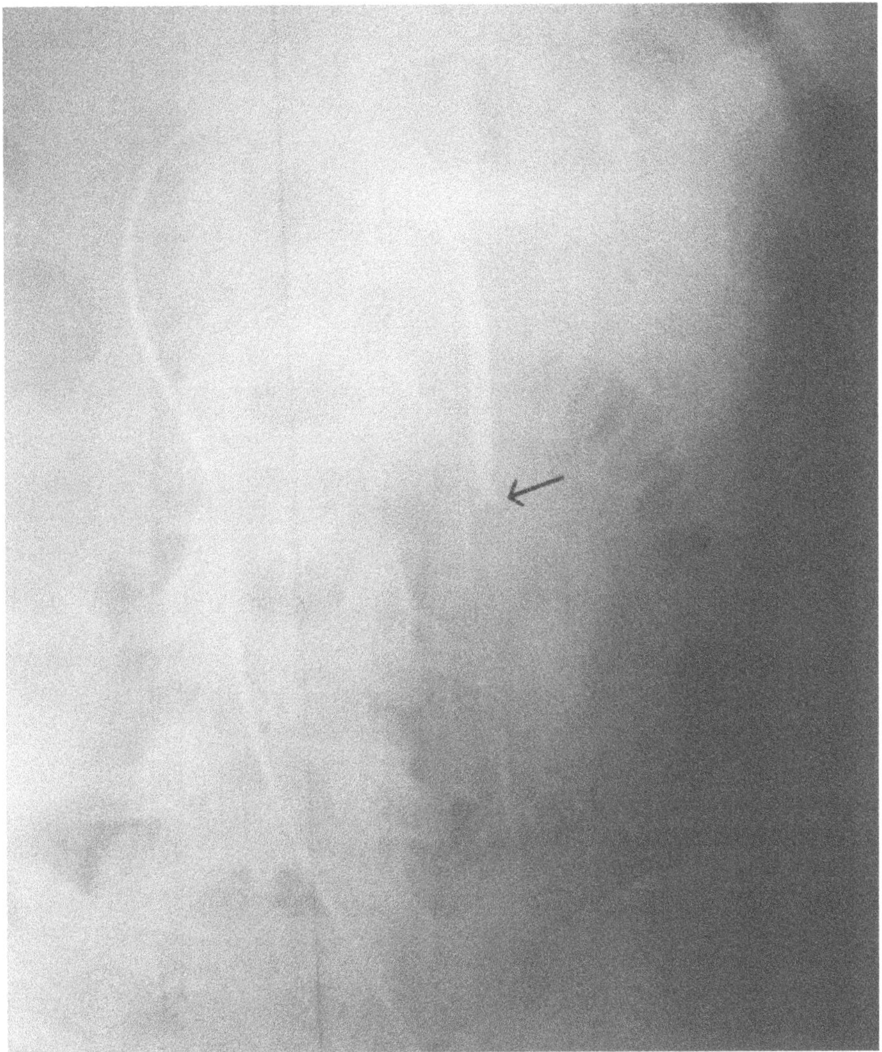

Fig. 70. A competent valve at a more caudal site in the internal spermatic
vein (↑) is bypassed by a delicate collateral vessel

B. Significance of Venous Valves for Development and Therapy of Varicose Veins

We distinguish between primary and secondary varices. *Primary varices* develop in patients with healthy and undamaged deep leg veins. During walking, the blood in the saphenous veins flows from the central to the peripheral region (Fig. 72 A). In contrast, *secondary*

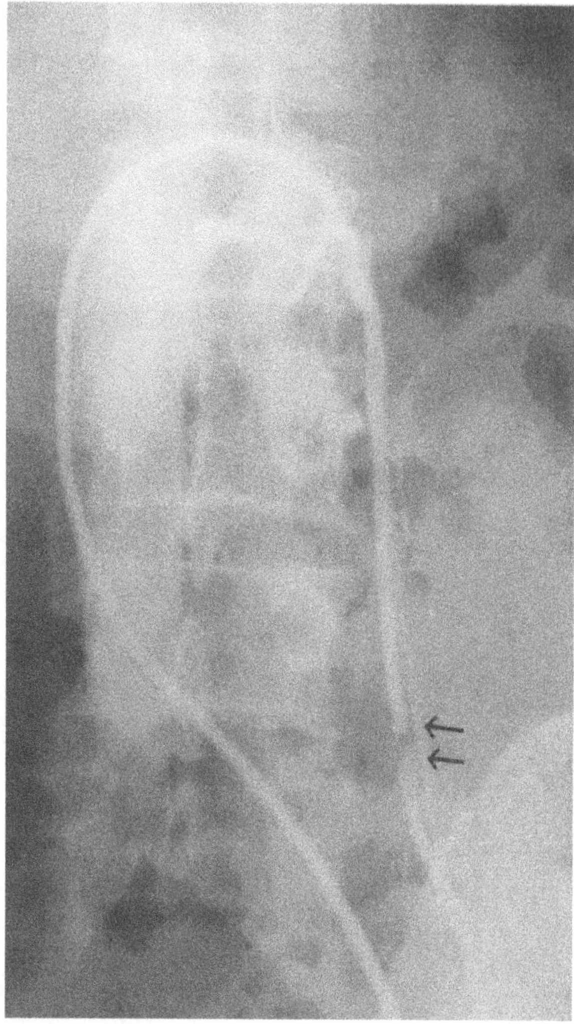

Fig. 71. A competent valve in the mid portion of the spermatic vein is bypassed by a delicate collateral vessel (↑↑). 1st degree varicocele

varices primarily involve an obstacle to the bloodflow in the deep leg veins. The superficial varices help out as collateral vessels. In doing so, they dilate and thereby also develop valvular incompetence. The decisive factor is that in secondary varices the bloodflow is mainly centralward as the vessels substitute only the damaged deep leg veins (Fig. 72 B). We shall now discuss the question of how primary varices develop. The development of varices has been explained by two different theories, some parts of which are diametrically opposed and which have had certain influences upon the treatment of varicosis.

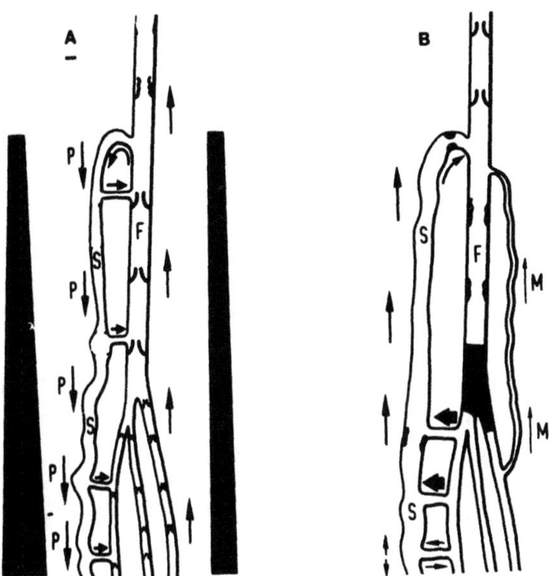

Fig. 72. The difference between *primary* and *secondary* varices, e.g. with post-thrombotic syndrome. **A** Primary varices. *S* long saphenous vein. *F* femoral vein. *P* private circulation. The arrows indicate the direction of bloodflow. The wedged columns symbolize the pressure in the superficial venous system (left) and in the deep venous system (right). The direction of flow is dependent on the difference of pressure. **B** Post-thrombotic syndrome. Note the changed direction of flow in the perforating branches and the varicose veins, which – in collaboration with the muscular branch M – have taken over the collateral circulation for the obliterated femoral vein and are therefore dilated

a) Theory of Private Circulation

As early as 1890, *Trendelenburg* referred to a private circulation of varices. According to this theory, the blood flows centralward in the deep veins but, due to incompetence of the valves of the long or short saphenous vein, some part of it enters the superficial venous system, where it flows off into the direction of the periphery in accordance with the force of gravitation. Through the communicating and perforating branches it rearrives in the deep venous system. In that manner some portion of the blood will keep recirculating – hence "private" circulation. This theory was further corroborated by *Perthes* (1895), one of *Trendelenburg's* collaborators. It has relatively few supporters today.

b) The Theory of Perforators

This theory was propagated mainly by *Linton* (1938). *Linton* observed that with primary varicosis the perforating and communicating branches were usually dilated and their valves were incompetent. In principle, the Theory of Perforators says that the venous valves of perforating or communicating branches – i.e. those vessels that connect the deep veins to the superficial veins – become incompetent. When the muscle pump tightens, the deep veins are compressed and, due to valvular incompetence, the blood may reach the superficial veins and inflate them, causing "blow out", which – according to the supporters of the Theory of Perforators – is an important factor in the development of varicosis. The Theory of Perforators is accepted by most authors today (*May, Partsch* and *Staubesand,* 1981).

c) Arguments in Favour of the Theory of Perforators

1. *"Blow Out"*

If the leg of a patient suffering from varicosis is wrapped up by rubber bandages, proceeding from the periphery in central direction, and the elastic compression is subsequently removed by unwrapping the leg in a downward direction to the periphery, certain areas will be visibly inflated in aneurysmatic venous dilations. One such "blow out" is frequently found approximately at the height of *Hunter's* canal. Other "blow outs" may be found at more distal positions.

2. *Phlebography*

If contrast medium is injected into varicose veins, the radiopaque matter will flow off into the deep veins through communicating branches. Conversely, filling of superficial veins with contrast medium may be achieved via the deep veins occasionally.

3. *Inspection of Varicose Leg*

When inspecting an upright standing patient with varicosis, we – in many cases – find varicose dilations merely in the periphery of the extremity, frequently beginning with the aneurism at the height of *Hunter's* canal mentioned above (Fig. 73). In rarer cases, visible venous dilations are also found at more proximal locations in the thigh.

4. *Surgical Findings*

As a rule, when operating varicose veins, one finds a superficial varix connected to a deep vein by an enlarged communicating vessel.

5. Mechanism of Recurrence

Recurrence after treatment of varicosis frequently occurs. Various authors have seen the causes of such recurrence in overlooked incompetent communicating or perforating branches.

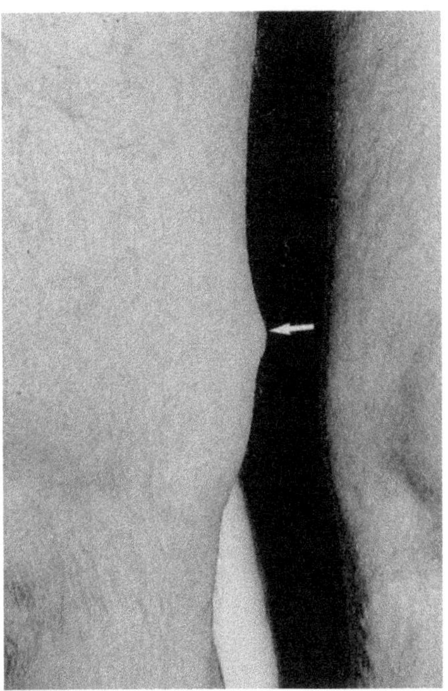

Fig. 73. Typical aneurismatic venous dilation approximately at the level of Hunter's canal (arrow)

6. Frequency of Perforating Branches

The total number of perforating veins in the entire leg is estimated to be 60. They have been classed under groups, such as *Dodd's* group in the thigh, *Boyd's* perforators below the knee joint, *Cockett's* perforators at the distal lower leg and *May's* perforators in the malleolar region. – According to the supporters of the Theory of Perforators, this multitude of perforating veins is responsible for the variegated picture of primary varicosis.

d) Arguments in Favour of the Theory of Private Circulation

Some arguments speaking in favour of the Theory of Private Circulation are given below.

1. Trendelenburg's Test

Trendelenburg's Theory of Private Circulation is based on the following observation: With the patient in supine position, the varicose extremity is lifted, blood is stroked out of the veins. The sapheno-femoral junction is compressed manually. If the patient is

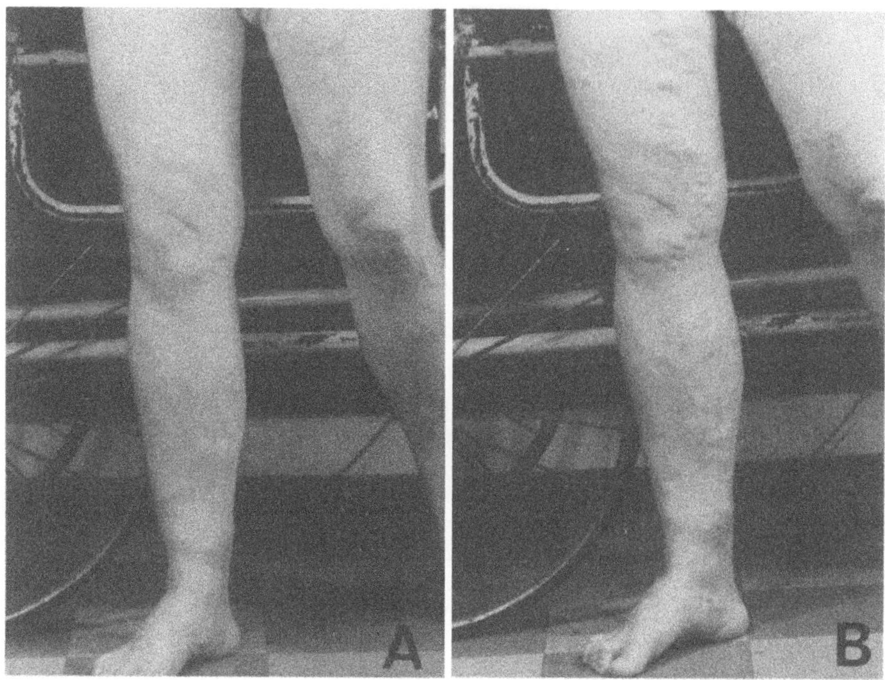

Fig. 74. Trendelenburg's test applied to varicose veins of the thigh and lower leg. **A** After drainage of the venous system in the lifted leg, compression of internal saphenous vein in inguinal region. After rising and putting weight on the leg, no filling of varicose veins. **B** After release of the compression, quick refilling of the veins all the way down to the foot

told to stand up now, the varicose veins refill only slowly and do so from distally to proximally. If the compression of the sapheno-femoral junction is suddenly released, there is a retrograde wave of blood, clearly visible in skinny patients, and the veins refill from proximally to distally, quickly attaining a state of maximum filling (Figs. 74 and 75). Carried out on patients with primary varicosis, this test is almost invariably positive. I found this confirmed in 200 consecutive cases among my own material. (See also *Edwards* and *Edwards,* 1940.)

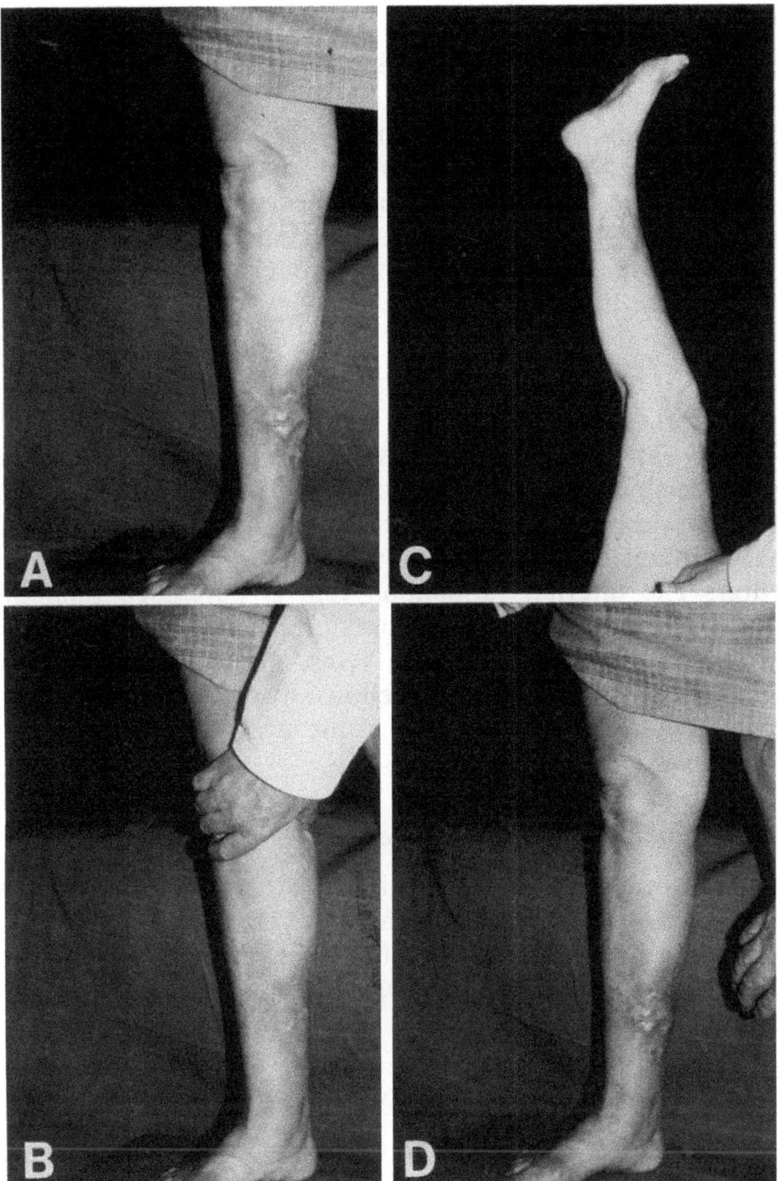

Fig. 75. Trendelenburg's test applied to a patient with varicose veins and marked pigmentation in the medial supramalleolar region of the right lower leg. **A** Inspecting the standing patient, one would feel tempted to suspect incompetence of Cockett's perforators. **C** The leg is lifted, the venous system stroked out. The internal saphenous vein is compressed at knee-level. **B** After standing up of patient, no refilling of the varices via the deep veins. **D** After release of the compression of the saphena, distinct refilling of the varicose veins, which may be misinterpreted as "blow out"

2. Perthes' Test

Observations of varicose-vein patients reveal that during standing the varicose veins are usually well filled. If the patient moves one may notice that during walking the filling of the varicose veins decreases. In order to prevent retrograde bloodflow in the veins, *Perthes* applied a rubber tube in a circular manner at knee level so that the superficial venous system was compressed at this site. The drainage of the varicose veins on the distal side of the rubber tube was very distinctly observed under these conditions.

3. Schwarz's Test and Reversed Schwarz's Test

Schwarz's test is applied as follows: When a varix is percussed very quickly or a varicose convolution is compressed manually, one feels – using the second hand – a wave of blood in the main trunk of the saphenous vein palpating proximally, occasionally even reaching up to the groin region. The test may also be applied in the opposite direction: The internal saphenous vein is percussed immediately distally to the sapheno-femoral junction, and a retrograde wave of blood is palpable in the varicose veins. – The test is indicative of incompetence of the main saphenous trunk, which need not be extremely enlarged. Its indicative value is roughly equal to that of *Trendelenburg's* test.

4. Retrograde Permeability to Probes

Usually, during operations of patients with widespread varicosis, a stripper inserted into the long saphenous vein via the inguinal region can be pushed forward far into the lower leg. In the presence of unimpaired venous valves this would be impossible without exertion of considerable force.

5. Pressure and Flow Measurements

Ludbrook and *Beale* (1962) proved valvular incompetence in the region of the veins above the sapheno-femoral junction of patients with varicose veins. Catheters were inserted into the iliac and femoral veins of patients or subjects afflicted with a hereditary taint. During *Valsalva's* maneuver as well as during coughing, retrograde pressure waves occurred, which were damped in healthy subjects but advanced undamped to the periphery in patients or subjects with hereditary taint. Investigations seem to indicate that in varicosity patients the valves proximal to the sapheno-femoral junction and in the iliac veins are missing so that intact venous valves are often found to be absent even in seemingly phlebologically healthy offspring of varicosity patients.

Pressure measurements were also carried out by *Recek* (1971) and *Recek* and *Koudelka* (1979). Doppler flow-measurements were made by *Folse* (1970), *Reagan* and *Folse* (1971), *Fegan* and *Kline* (1972), and *Schlunk* and *Göltner* (1973). *Wuppermann* (1981) believes that Doppler-ultrasound examination is the most accurate method for evaluating valvular incompetence in superficial veins. The studies of the above authors may be summed up by stating that as a rule in varicose veins venous valves in the region of the proximal saphenous vein as well as in more central regions are incompetent.

Fegan and *Kline* (1972) found turbulences in superficial leg veins after increased pressure in abdominal veins only under conditions of primary varicosis. This finding also supports the assumption that functional valves are missing.

Tibbs and *Fletcher* (1983) reported on flow measurements in superficial veins by means of the directional Doppler instrument. Obviously unaware of *Bjordal's* results described in the following paragraph, they conclude that in standing patients with primary varicosis the bloodflow in the saphenous vein is in peripheral direction, whereas in obliterated deep veins it is cranialward, and oscillates in both directions under conditions of valvular incompetence of the superficial and deep systems.

6. Flow Measurements in Exposed Veins

By means of flow-meters, *Bjordal* (1970–1972) measured the flow in varices of communicating branches and of the main saphenous trunk. *Bjordal* fully confirmed *Trendelenburg's* statement according to which there is retrograde flow in the main saphenous trunk of patients with varicosis. In the communicating branches there is outward flow during the muscular systole and inward flow during the muscular diastole, with the overall inflowing quantity of blood being in excess of the outflowing quantity. According to *Bjordal,* in walking patients the venous pressure at ankle height could be reduced by 17 mm Hg when the long saphenous vein was compressed at a more proximal site, whereas the compression of a perforating vessel resulted in a pressure reduction of no more than 2 mm Hg. – These investigations also support the Theory of Private Circulation. *Bolliger* (1981) came to similar conclusions by measuring the flow in perforating veins by means of introduced ultrasound flow-meters.

7. *Morphologic Findings in Venous Valves*

On the basis of morphologic examinations, *Basmajian* (1952) does *not* believe that primary valvular incompetence is the cause of varicosity. The morphologic investigations conducted by *Leu et al.* (1979) have led to similar conclusions. They did *not* find regression of venous valves with advancing age, and in many cases the uppermost valve of the long saphenous vein, slightly before the junction of the vessel with the femoral vein, was competent. From these findings the authors inferred that dilation of the vein was not caused by primary loss of valves but that primarily there was loss of venous tonus. Valvular incompetence, if any, was secondary.

Investigations of our own (*Gottlob et al.,* 1975) using another method (en-face preparations, silver staining) have shown other results. Out of 22 uppermost valves of the long saphenous veins of varicosity patients, a normal cusp was found in only 5 cases (Table 2). Even in these cases valvular incompetence could not be outruled due to the fact that during the process of cutting the veins open the respective partner-cusps were damaged. There could also have been incompetence caused by dilation of the vein without any pathologic changes visible on the cusp. – These findings were discussed in detail on p. 52.

Results of Surgery and Causes of Recurrences

Surgical therapy of varicosis neglecting the communicating and perforation branches and involving merely crossectomy and stripping of the long saphenous vein while sclerosing its branches, e. g. by injection therapy, does not result in enhanced frequency of recurrences as compard to extensive ligature of perforators. At closer examination, the recurrences observed by one of the authors (R. G.) – with one single exception – were to be attributed to the development of a fine network of vessels originating from the femoral vein below the inguinal ligament and, in most cases, uniting at a confluence in the form of a somewhat larger vessel located distal in the thigh. – This tendency toward recurrence is explained by increased pressure conditions prevalent in this region. These increased pressure conditions are due to the fact that as a rule valves are absent at more proximal sites, so that, as a consequence of the increased pressure, collateral vessels develop and find the entrance to a smaller vessel of the thigh, which subsequently dilates. – Numerous experimental studies have demonstrated the enormous plastic capacity of the venous system.

Figs. 76 A and B show the mechanism of recurrence occurring in varicosities operated or sclerosed without high ligature of the long saphenous vein. In the remaining blind sac of the saphenous vein there is a considerable rise of pressure and an increased gradient of pressure towards the preserved peripheral branches. This gradient of pressure is responsible for the dilation of collateral veins. – Fig. 76 C contains a schematic illustration of the mechanism of recurrence after high saphenous ligature. Fig. 77 shows a corresponding X-ray film.

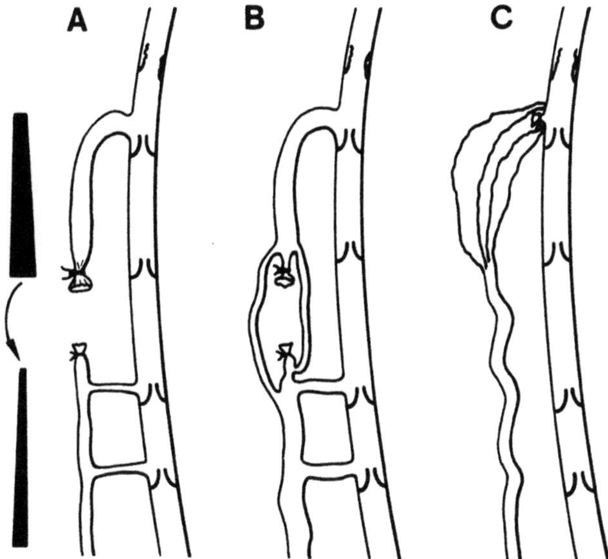

Fig. 76. **A** and **B** Mechanism of recurrence after non-radical surgery of varicose veins (refraining from high saphenous ligature). The wedged black columns symbolize the pressure conditions, the curved arrow indicates the gradient of pressure. **C** Schematic illustration of mechanism of recurrence after high saphenous ligature

e) Critical Discussion of the Two Theories Explaining the Etiology of Primary Varices

Before comparing the arguments brought forward by supporters of one of the two theories, one should not fail to make the following remarks for the sake of fairness:

1. The overwhelming majority of authors adheres to the Theory of Perforators. Besides *Trendelenburg* and *Perthes, Recek* (1971) and *Recek* and *Koudelka* (1979), as well as with some reservations *Bjordal*

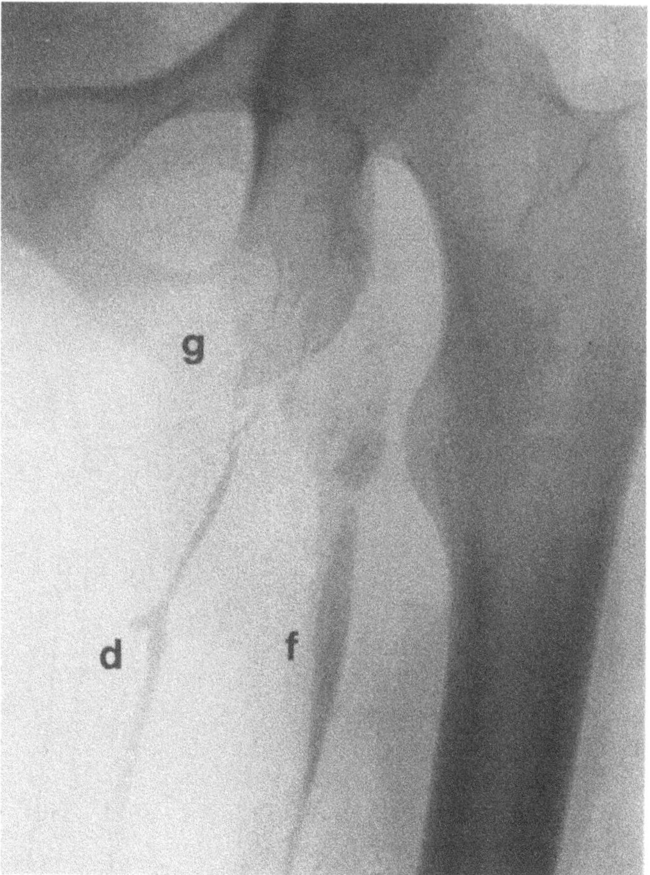

Fig. 77. Recurrence of varicosis after correct high saphenous ligature and stripping. *d* duplicature of long saphenous vein. *f* femoral vein. *g* network consisting of several fine veins, connecting the femoral veins to the saphenous duplicature

(1970, 1972 and 1983), argued for the Theory of Private Circulation in more recent years.

2. The author of these investigations has come out for the Theory of Private Circulation in a number of publications (*Gottlob*, 1968, 1977, 1978, 1981). For this reason, he might be biassed in this question and it must be left to the reader's discretion to critically weigh his arguments.

3. For the sake of completeness it must be mentioned that there are many other theories concerning the etiology of varices, which are not based on the destruction of valves, such as the theory of arteriovenous anastomoses, propagated by *Piulachs* and *Vidal Barraquer*

(1953). Other theories include that of compression of the femoral vein in the lacuna vasorum (*Gullmo,* 1964) or that of venous dilation due to intraabdominal pressure caused by straining during defecation (*Burkitt,* 1972). – These theories have not received widespread acceptance either.

4. Finally, it must be mentioned that even the two authors of this book do not fully agree in their views concerning the genesis of primary varicosis so that the following critical review is based mainly on the view of one of the authors (R. G.).

1. Visual Impression

For the following reasons, the visual impression as it presents itself during inspection of patients with varicosis may lead to the assumption that perforating or communicating branches are involved. When inspecting a standing patient, one often notices that the varices do not originate from the inguinal region but only from the distal thigh or occasionally even from the lower leg downward. This is due to the following causes:

In the thigh, the long saphenous vein lies in the depth under a thick pad of fat, which is of particular thickness in some women. Therefore, it is evasive of inspection and frequently also of palpation. Besides, in the proximal thigh the long saphenous vein is covered by a thin duplicature of fascia, which was described by *Sherman* (1948). *Kubik* and *May* (1979) do not refer to a duplicature of fascia but to a subcutaneous pseudo-fascia, covering the venous trunks and lymph nodes. Because of this fascia or pseudo-fascia, the long saphenous vein of the proximal thigh is often prevented from more extensive dilation although the vessel is already incompetent.

Besides, the hydrostatic pressure in the distal portions of a standing patient is higher than in the proximal portions so that the veins in the periphery have a stronger tendency to dilate than veins in more central locations, which may already be suffering from valvular incompetence without showing marked signs of varicosis.

2. "Blow Out"

This phenomenon seems to induce an observer in a particularly deceptive way to the assumption that perforators do play an important role in the genesis of varicosis. Fig. 75 shows a *Trendelenburg's* test. In Fig. 75 A an extremity is lifted and compressed after stroking out the blood from the superficial veins. When the patient rises, while the compression of the saphena at knee level is maintained, visible varices in the region of *Cockett's* perforators fill up slowly and under modest pressure. If, however, the compression of the saphenous vein is released, there is a retrograde wave in the long saphenous vein and the varicose veins fill immediately. Even compression of the long or short saphenous vein slightly above the "blow out" suffices to show that the entire phenomenon is misleading. In all cases of primary varicosis it may be proved that the varices are not filled from the deep regions but from proximal regions via the main saphenous trunk. This observation raises doubts on the arguments of the advocates of the Theory of Perforators. The "blow out" is primarily a "blow down". Secondarily, one might speak of a "blow in" as the overall flow, e.g. during walking, is directed toward the inside. The term "blow out" is justified in third consequence only. It is effective only in the moment of muscular tension and the quantities of blood transported to the surface are considerably lower than those transported from outside to inside.

3. Dow's Sign

Dow's sign refers to the bulging out of a superficial vein at the entrance of an incompetent perforating vein (*Dow,* 1951). *Dow's* sign is also frequently stated in support of the Theory of Perforators. In our opinion it is no more conclusive than "blow out" and based on the same principle, i.e. filling of the valve pocket in a downward direction. The mechanism causing *Dow's* sign is to be explained as follows:

As a rule venous valves are located slightly distal to an entrance into another vein. Therefore, the perforating veins practically discharge into the valvular sinus. The region of the valvular sinus is the thinnest portion of the venous wall, as shown in Fig. 10 A. If, due to incompetence of the main saphenous trunk, a vein is under increased hydrostatic pressure, dilation is possible, particularly in the region of the valvular sinus where the venous wall is thinnest. Therefore, *Dow's* sign is caused not by enhanced outflow from the deep system but by increased downward pressure. The fact that perforators empty at these very sites is to be attributed to the pattern of valvular arrangement near entrances.

Cotton (1961) published findings partly contradicting the above interpretation. According to *Cotton,* varices in the saphenous region occur *distal* to the valvular cusps, i. e. *not* in the region of the valvular sinus having a more proximal location. The varices are found invariably on one side under a valvular cusp; the sinus under the second cusp may be enlarged but never dilated to a larger varix. – This is also evidence against the Theory of Perforators as in this location the aneurismatic varix is protected from the bloodflow entering from the perforator proximally to the valvular cusp.

The location of the varix distal to a venous valve as reported by *Cotton* confirms older findings by *Trendelenburg* (1880) and *Slawinsky* (quoted according to *Hasebrock,* 1916).

On the basis of extensive model experiments *Hasebrock* concludes that the aneurisms distal from valves are caused by pulsation transferred to veins by the accompanying arteries. The following objection to this hypothesis must be pointed out. We know today that under conditions of primary varicosis the bloodflow in the dilated long saphenous vein is effected mainly in the opposite direction, i.e. the flow is proximal to distal. Moreover, it must be stated that the pulsation transferred to the venous system by the arteries is not provable in superficial veins by means of pressure measurements.

It is difficult to find an explanation for the development of aneurismatic varices distal to venous valves. One explanation, which is merely of hypothetical character, might be based on the well-known principle of post-stenotic dilation. Post-stenotic dilations are observed mainly in the arterial circulation. In varicose saphenous veins there is a strong retrograde current, i.e. "private circulation", in particular during walking. Valvular aggers represent the thickest areas of the venous wall. They exert more resistance to increased intravasal pressure than do other regions of the venous wall so that when the venous trunk is dilated, the region of the agger maintains its original diameter the longest. What develops here is relative stenosis, and the area distal to the valve is now subjected to post-stenotic dilation. Due to the considerable resistance to retrograde flow exerted by the valves or their rudiments, we may expect that in this region there will also be marked turbulences, which may also contribute to dilation.

The assertion according to which venous dilation in the form of aneurismatic varix is to be found peripheral to venous valves is in contradiction to our own previously held view. Therefore, we checked on the assertion in a number of cases and found it supported in most of them (Fig. 78). Our findings showed even in this case that one cannot lay down a rule without having an exception: in some cases aneurismatic dilations were found in the region of the valvular sinus. So, once again, we are not dealing with a "law" that holds true in every case.

4. Phlebography

While as a rule phlebography may provide satisfactory information with respect to the anatomic situation, it does not provide indicative information regarding the pathogenesis. If contrast medium is injected into a superficial vein or a varicose vein, it flows off into the deep veins via the perforators, following the physiologic direction of

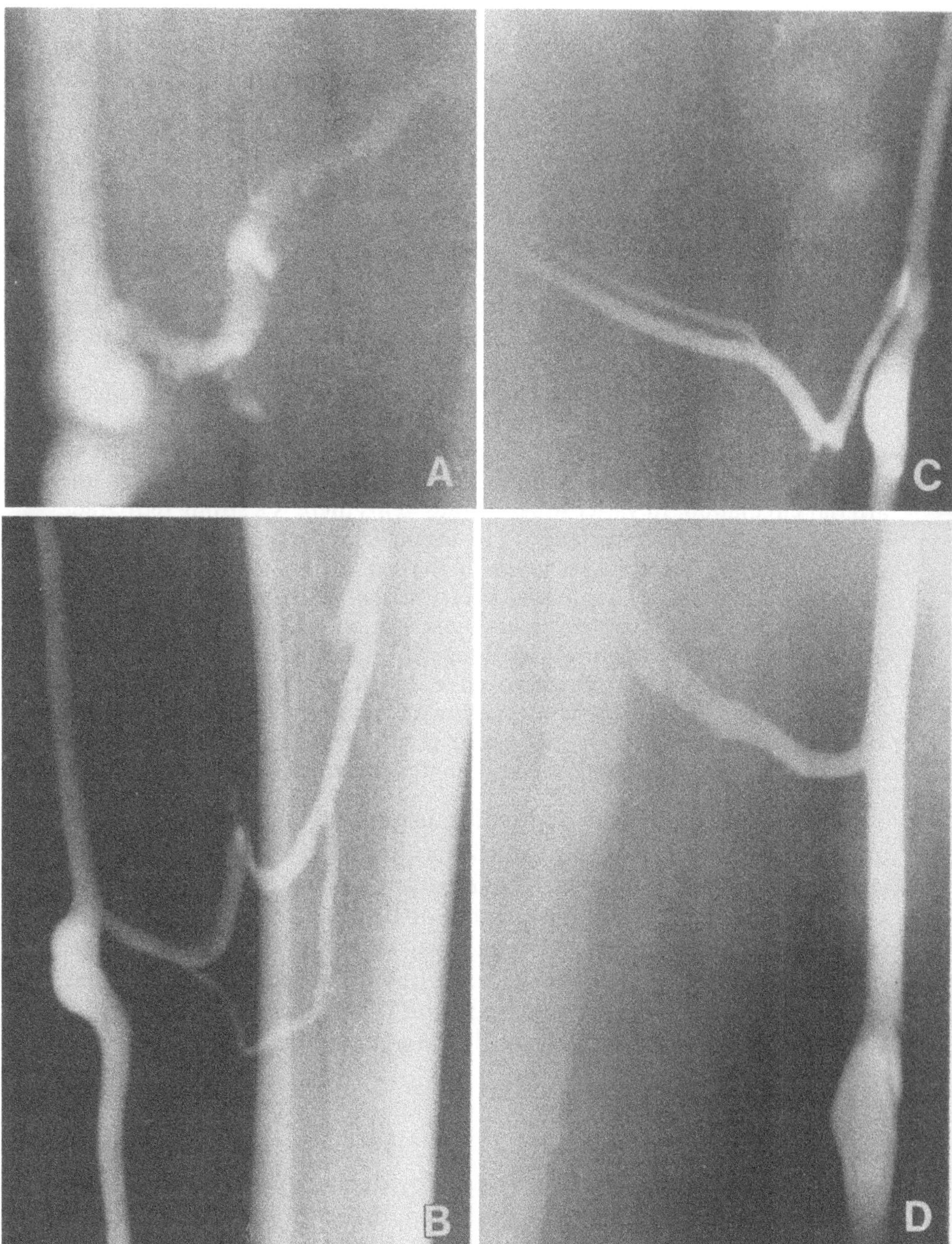

the bloodflow. Therefore, visualization of the perforating veins is by no means conclusive of the pathogenetic importance of these vessels. – If the contrast medium is injected into the deep veins, occasionally perforating veins are visualized, and superficial veins or varicose veins are filled via those perforators. But even in this direction of flow, the visualization of perforators is by no means conclusive. We know that radiologic contrast media, due to their higher specific weight, tend to sedimentation. Contrast media injected into the caval vein may sink as far down as into the veins of the lower leg despite the presence of intact valves. Due to the higher specific weight of the contrast media this occurs contrary to the bloodflow, while the valves are kept open by the bloodflow itself. Therefore, we could imagine that the contrast medium may also penetrate into the superficial veins from the deep veins via the perforators rather than rise upward with the current in the deep veins. – At any rate, it is conspicuous that by filling of the deep veins, e. g. ensured by application of a tourniquet above the point of injection, it is usually impossible to visualize varicose veins over their entire length. No more than fractions of the macroscopically discernible varices are shown. In view of the hydrodynamic properties of contrast medium, the specific weight of which is higher, incompetence of a communicating vein should be diagnosed with utmost caution. But even positive evidence of such incompetence will not be indicative as to the pathogenesis of varicosis. After all, it is not an unlikely assumption that the communicating veins of peripheral leg veins might become expanded in the same manner as peripheral leg veins dilate under the high hydrostatic pressure caused by incompetence of the main saphenous trunk. The likelihood of the assumption seems even greater when one takes into consideration the substantial increase of bloodflow (described in the next subparagraph) observed in primary varices in the region of the communicating veins.

5. Flow Measurements

Retrograde bloodflow in the region of the main saphenous trunk under conditions of primary varices was proved by *Folse* (1970),

---

Fig. 78. Dow's sign. **A** A perforating vein entering into a valvular sinus. The vein is also dilated distal to the sinus. **B** Incompetent Boyd's perforating vein enters above a valve. The dilation is mainly distal to the valve. **C** Similar situation. The perforating vein empties into a valvular sinus, distal to which the vein is irregularly dilated in a spindle-shaped manner. **D** Spindle-shaped dilation distal to a valve but the perforating vein enters at a much more cranial site

Barnes et al. (1975) and *Wuppermann et al.* (1981). *Reagan* and *Folse* (1971) showed that incompetent ileofemoral valves were found in 16% of normal subjects, whereas among the offspring of varicosity patients the percentage of such incompetence was 32.

Flow measurements by an electromagnetic flowmeter were carried out by *Bjordal* (1970, 1972, 1981 und 1983). *Bjordal* applied the flowmeter to both superficial veins, especially over the main saphenous trunk, and perforating branches. Pressure measurements readings were obtained simultaneously. A schematic illustration of *Bjordal's* results is given in Fig. 72. *Bjordal* arrives at the following conclusions: The pressure in primary varicose veins at rest is fairly constant. When the leg is relaxed and at rest, the flow in the varices and the perforators is almost completely suspended. Only in the deep veins is there centralward flow. During walking, there is retrograde flow in the varicose veins, the volumes per minute of which were observed to be at levels between 175 and 500 ml p. min. When the subject is walking, there is outward flow during the muscular systole and inward flow during the muscular diastole in the perforating branches. Due to the strength of the inward flow during the diastole, the resulting overall flow is directed from the outside to the inside! – *Bjordal* thus verified the private circulation asserted by *Trendelenburg* by an exact method. He writes (quoted verbatim):
"According to the current view, the pathological ambulatory venous hypertension found at the ankle in the superficial veins in the presence of primary varicose veins with dilated, incompetent calf perforators, is transmitted through these perforating veins. This concept strongly conflicts with the observations of the present investigation, which indicated that venous hypertension during ambulation is caused by the retrograde flow through the saphenous main channel and is not transmitted through the dilated perforators."

It is difficult to find an explanation for one of *Bjordal's* findings according to which there was a considerable increase of the bloodflow from the deep system to the surface when the long saphenous vein was compressed proximal to the perforators. *Schneider* conducted Doppler-ultrasound examinations under similar conditions. The increased bloodflow to the surface might be explained by arguing that when the long saphenous vein is compressed at a central site the pressure in the saphena is reduced so that the flow from the deep to the superficial veins may increase.

6. Pressure Measurements

In addition to those conducted by *Bjordal,* pressure measurements were also carried out by *Ludbrook* and *Beale* (1962), *Recek* (1971) and *Recek* and *Koudelka* (1979). These investigations also resulted in indications of valvular incompetence, in particular in the region of the common femoral vein and the region of the iliac veins.

7. Mechanism of Recurrence

If a stump of the long saphenous vein is left in the inguinal region or a larger branch is not ligated, recurrence at the respective site will most probably occur. This is due to the facts that below the crossectomy there is lower pressure in the peripheral superficial veins as the connection to the central veins has been broken, and the pressure in the deep veins is decreased by the functional valves. In contrast thereto, there is relatively high pressure on the central side of the ligature. This is due to the fact that usually the valves above the sapheno-femoral junction are incompetent in patients with varicose veins. Thus, there is a steep pressure gradient by which the development of collateral vessels is stimulated. According to *Luke* (1954) and *May* and *Nissl* (1973), most cases of recurrent varicosities are to be attributed to incompetent perforating branches. On the other hand, a summary by *Pouliadis et al.* gives 42 high-type recidivisms out of 48 recurrent cases and no more than 6 for which perforating branches were held responsible. Out of 204 recurrences of varicosities, *Sheppard* found recurrence of sapheno-femoral incompetence in 90% and exclusive incompetence of perforators in merely 1%. Similar ratios were found by *Gottlob* in material of his own; most of the recurrences were caused by the formation of a network of thin vessels between femoral vein and a superficial vessel in the fossa ovalis. Occasionally, this superficial vessel was a duplicature of the internal saphenous vein. Alternatively, it was another subcutaneous vessel (Fig. 77).

The varying mechanisms of recurrence as described by the authors may be attributed to the varying operational techniques employed. Another explanation would be that while the perforating branches were present, there were additional connections to the femoral vein in the region of the fossa ovalis. These connections evaded phlebographic visualization due to the reasons described above.

8. Destruction of Venous Valves

According to the Theory of Private Circulation, incompetence of the main saphenous trunks is caused by destruction or at least incompetence of the valves in this region.

According to statements dating back to earlier periods, the number of venous valves in adults is only a fraction of that found in fetuses or newborn infants. Several more recent authors failed to find those statements verifiable, e.g. *Kampmeier* and *La Fleur Birch* (1917) and *Leu et al.* (1979). However, their findings conflict with the above-mentioned functional studies (Doppler-ultrasound and pressure measurements) by *Ludbrook* and *Beale* (1962), *Folse* (1970), *Reagan* and *Folse* (1971), *Recek* (1971) and *Recek* and *Koudelka* (1975).

Animal experiments of our own have shown that valvular incompetence will usually develop after dissolution of thrombi in the valvular region. This valvular incompetence is occasionally combined with considerable shrinkage of the valvular cusps. While *Leu et al.* found the so-called main valve (terminal valve), i.e. the uppermost valve of the long saphenous vein, to be present in varicosity patients in all cases, investigations of our own (*Gottlob et al.,* 1975) have revealed changes of the uppermost valve of the long saphenous vein in a vast majority of varicosity patients. These findings were discussed elsewhere in this book (pp. 110ff.). Here we would confine ourselves to pointing out that the morphologic evaluation of valvular incompetence may be extremely difficult at times. If veins are dilated, valves may be incompetent without showing any morphologic changes. This is simply due to the fact that the cusps are too short for the extended diameter of the lumen. Besides, for the purpose of evaluation of the valves, the veins must be cut open. In this course one cusp is destroyed as a rule and thus evaluation of this valve is prevented. Histologic sections offer views of only tiny fractions of the valvular area, and localized adhesions or minor-degree shrinkage often slip by unnoticed, unless serial sections showing the entire valvular area are made. The frequent occurrence of valve-pocket thrombosis was emphasized by *Sevitt* (1974). *Sevitt* describes 50 small-sized thrombi in valvular sinuses of femoral veins and discusses their significance as preliminary stages of more comprehensive thromboses. On the basis of experimental studies of our own we assume that, in the event of organization, such thromboses result in shrinkage and subsequent disappearance of venous valves, leaving behind only more or less visible rudiments. If the deposition of thrombi in the region of the valvular sinus is only of minor nature, a chain reaction may develop in that, for the time being, the cusp will shrink only moderately, thus becoming slightly incompetent. Due to the regurgitation of blood caused by incomplete closing of the valve, there will be turbulences, promoting the deposition of new microthrombi, which in turn may cause progressive shrinking and subsequent disappearance of valves. – For the causes of the destruction of venous valves see p. 23.

f) Circulatory Effects of Varicose Veins

Minor-degree varicosities usually are no more than cosmetic impairments to the patient. If the varicosities are more extensive, the increased volume of the vessel and absent venous valves cause shifting of considerable quantities of blood, e.g. when the patient changes his position from lying to standing. *Arenander* (1960) summed up his findings obtained by circulatory examinations like this: On rising, the patient undergoes a blood withdrawal and on lying down he receives a blood transfusion. In the patients examined by *Arenander* these shiftings of quantities of blood in orthostasis resulted in higher increases of pulse rates than in healthy persons. In orthostasis both pulse pressure and heart volume were reduced.

In the affected extremity the bloodflow through the deep leg veins is increased due to the fact that these vessels must transport both the blood necessary for the nutrition of the leg and, in addition, the blood of the "private circulation". In extreme cases even the deep veins may become incompetent in this manner. As a consequence, "mixed dysfunction" (*Bjordal,* 1982) may develop.

Other Sequelae of Primary Varicosis

Under conditions of severe varicosis we find phenomena similar to those encountered in the post-thrombotic syndrome, i.e. itching of the skin in the region of the distal lower leg, indurations, pigmentations and ulcers.

Arnoldi and *Haeger* (1967) deny the occurrence of ulcers in primary varicosis. *Hoare et al.* (1982), on the other hand, report on 20 thoroughly examined patients with primary varicosis, who were found to suffer from a total of 23 crural ulcers. Incompetent perforating veins were found in only 14 of these 20 patients!

Do varicose veins cause *cramps?* The german word "Krampfadern" would suggest such an association. *Santler,* however explains this misleading expression by a popular etymology (1974).

g) Localization of Communicating Branches with Valvular Incompetence

So far, several methods have been suggested for localizing incompetent communicating branches.

Clinically, Otto (1963), and *Wuppermann* (1973) palpated the gaps in the fascia through which the perforating veins connect to the superficial veins. The quota of hits is estimated between 50 and 75% (*Netzer,* 1979).

Ultrasound Diagnosis. This method was applied by *Folse* (1970) and *Spiess* (1972). The method is encumbered by the difficulty of eliminating stray noises. According to *Wuppermann* (1973), the quota of hits was 85% while in 15% erroneous localizations were obtained.

Thermography. This method was applied by *Beesley* and *Fegan* (1970), *Patil et al.* (1970), *Elem et al.* (1971), *Noble* and *Gunn* (1972) and *Vuori et al.* (1972). By lifting of the extremity, the superficial veins of the leg are drained. Two tourniquets serve to prevent refilling of these superficial veins by other superficial veins at more proximal or distal locations. Then the leg is lowered and cooled by ice. The patient is told to move the forefoot. After an interval of approximately one minute, "hot spots" are noticeable at the points of exit of the perforating veins. The quota of hits ranges between 15 and 90%. Erroneous localization is obtained in 15%.

Fluorescence Phlebography. A 5% solution of sodium-fluorescein is injected into a vein in the back of the foot. In a similar manner as in thermography, the point is ascertained at which the blood passes from the deep veins into the superficial venous system. In the darkened room, these points glow under ultraviolet light. *Chilvers* and *Thomas* (1979) report a quota of hits of 90%. *Elem et al.* (1971) and *Noble* and *Gunn* (1972) give considerably lower quotas of hits (16 and 47% respectively).

Phlebography. This method is the most accurate of all single methods. We will describe the phlebographic method in detail later on.

Multiple Methods. By a combination of various of the above-mentioned methods, in particular by a combination of clinical examination (palpation) with radiology or other methods described, quotas of hits of almost 100% may be achieved.

Table 4. *A comparison of various methods of localization of insufficient perforating veins*

	Number of authors	Average quota of hits (%)	Lowest and highest quota of hits
Palpation	10	50	34–63
Ultrasound	6	75	55–95
Thermography	4	72	39–94
Fluorescent venography	6	57	16–96
Phlebography	7	59	44–69

(Calculated according to *Wienert*, 1981, from collective statistics.)

Table 4 compares the value of the methods of localizing perforating branches with incompetent valves.

h) Therapeutic Consequences

There is general unanimity that treatment of extended varicose veins should comprise high saphenous ligature. This applies to both long and short saphenous veins, as remaining blind sacs greatly add to the risk of recurrence and, besides, offer favourable conditions for the development of thrombi and embolism. There is also unanimity that the enlarged trunks of the vessels should be removed by "stripping", surgical excision or sclerotherapy. There is disagreement, though, with respect to the question of what should happen with the perforating or communicating branches.

It would definitely be a mistake to aim at extreme radicality. The number of perforating veins contained in a leg is estimated at 150, some 60 of which have lumina of considerable size (*Schäfer*, 1981). Methods have been developed with a view to eliminating at least a majority of these vessels. For this purpose it was necessary to perform extended longitudinal incisions, making any cosmetic improvements whatsoever impossible.

The alternative extreme is that of restricting therapeutic measures to high saphenous ligature, stripping and sclerotherapy, completely disregarding the perforators (*Gottlob*, 1981; *Recek*, 1971; *Sheppard*, 1978). These authors maintain that they have not observed worse results after having refrained from ligature of the perforators.

The middle of the road measure is to expose conspicuously enlarged perforators whose incompetence may be safely assumed and locally ligate them (*May*, 1981; *Bassi*, 1981). (The branches are exposed by means of a short transverse incision of the skin.) While it is difficult to objectify the necessity of this measure, one should not altogether reject the method, if only for its innocuousness. – Even after application of the hooking method (*Bassi*, 1973) recurrence has been observed.

In any case one should not fail to distinguish between primary and secondary varicosis in the treatment of varicose veins. Many authors seem to be unaware of the fundamental difference, which is shown in Fig. 72 (p. 129). *Hyde* and *Hull* (1981) performed subfascial ligatures of the perforating branches on 109 patients. As few as 31 patients had a past history of phlebitis. Among 83 followed-up patients, there were 27 cases of recurrent ulcers, and 19 patients required another operation. The authors conclude that ligature of the perforators is only a palliative measure. – We would add that, as a rule, treatment of ulcers in patients with primary varicose veins is relatively easy. After extirpation of the varicose veins and high saphenous ligature, the ulcer will readily heal.

Bjordal (1981) differentiates his procedure in that he does not ligate perforators in early stages. He assumes that they will regain competence after normalization of pressure conditions achieved by high saphenous ligature and stripping. In the case of longstanding and severe varicose veins with pencil-thick perforators, however, the valves in the region involved have been damaged definitely and irreversibly, and the vessels should therefore be removed.

Sclerosing therapy may be contemplated as a possibility for minor-degree varicosis and, in addition, for treating residual varicose veins, e. g. dilated peripheral branches persisting after surgery. According to experience of our own, the best results are obtained by surface-active substances. Solutions of single compounds are preferable to solutions of mixtures. The patient's vein is punctured while he is standing, and the contrast medium is injected while he is lying.

Several authors described the histologic changes occurring after injection of sclerosing solutions into veins, e. g. *Fischer* (1972).

i) Summary

Under physiologic conditions blood flows from the superficial veins through the perforating or communicating branches into the deep veins and, from there, centralward. In primary varicose veins, in particular during exercise, we observe a "private circulation", i.e. retrograde bloodflow in the long or short saphenous vein, by which the bloodflow in the deep veins is increased considerably. This may result in incompetence of the valves in perforating veins. In that case, a short wave of blood is thrust out into the superficial system during the muscular systole, and a flow of considerably greater intensity from the surface enters the deep veins during the diastole. The cause of primary varicosis seems to be valvular incompetence or absence of valves in the regions of the iliac veins and the veins above the sapheno-femoral junction. Under such conditions, the valves of the saphenous veins are subjected to increased hydrostatic pressure and the saphenous veins themselves dilate. This is due to the fact that they are not protected by surrounding striated muscles so that eventually their valves become progressively incompetent in a proximal to distal succession. This may be combined with shrinkage or complete regression of valves due to valve-pocket thrombosis.

For therapy of severe varicosis, high ligature of the long or short saphenous vein and removal of the dilated veins by stripping or sclerotherapy is recommended. At the present time unanimity has not been reached as to the extent to which perforating veins with incompetent valves should be ligated.

C. The Post-thrombotic Syndrome

The post-thrombotic syndrome is a widespread disease incapacitating a considerable percentage of the population or at least forcing them to change their occupation.

a) Clinical Symptoms

The post-thrombotic syndrome does not begin simultaneously with venous thrombosis. Instead, after the acute inflammatory phase, the patient passes through an interval of several months or even years, during which he is free of complaints. At the eventual onset of complaints, the patient suffers from edemata of the distal portions of the lower leg and of the ankle region. These edemata occur predominantly after prolonged standing and are more marked in the evening and less marked or even missing in the morning after bedrest. Without proper treatment changes of the skin such as itching, pigmentation, induration, and finally, exulceration will occur. The patient may suffer from severe ("bursting") pain. In advanced stages, the ulcers may become circular and involve large areas of skin. These ulcers are localized to the distal side of the lower leg, particularly in the supramalleolar region.

Lower leg ulcers occur in both primary and secondary varicosis. *Johnson et al.* (1985) found past histories involving thrombosis of the deep veins in no more than 23% of the cases of crural ulcers examined by them. A further 19% had suffered fractures of the leg or undergone orthopedic surgery prior to the occurrence of the ulcers. "So at most, 42% may have had deep vein thrombosis. More importantly, descending venography in 26 patients rarely identified recannulated veins. These observations suggest that there is a subgroup of patients with venous ulcers who have primary valvular incompetency." *Cockett* (1955) also found past histories of thrombosis of the deep leg veins in no more than 21% of his cases.

What is conspicuous about the cases described by *Johnson et al.* (1985), is the high rate of recurrence after ligature of perforating veins (22% after 1 year, 41% after 3 years, and 31% after 5 years).

b) Pathogenesis of the Post-thrombotic Syndrome

We must distinguish between obliterations of the leg veins and of the pelvic veins. Acute thrombosis of the *leg veins* results in inflammation and reduced blood supply, which may even cause "venous gang-rene". After the acute phase, there is occlusion of the deep veins, but the venous blood of the leg is drained via dilated collateral veins. In view of the fact that such collaterals may develop rapidly, the scarcity of vessels for the purpose of venous outflow is a problem of secondary nature. There will be no complaints before recannulation of the obliterated veins. Prior to that, thrombi in the deep veins cause dilation of the superficial veins, which are connected to the deep veins via enlarged perforating veins. Other than is the case with primary varicosis, the blood in *secondary varicose veins* flows centralward during walking, and in the perforating veins predominantly from the deep veins to the superficial veins. The difference of flow directions between primary and secondary varicosis is a decisive factor for diagnosis and therapy (see Fig. 72, p. 129).

During *recannulation* of the thrombosed portion of the deep leg veins, two pathogenetic mechanisms come into action:

1. The *muscle pump* is impaired in its function. Due to the absence of valves in the recannulated veins, the centralward direction of the bloodflow is not ensured. During muscular diastole, while the deep veins are dilated, some part of the ejected blood may flow back. The situation is similar to that of insufficiency of the aortic valve, where some part of the ejected blood regurgitates during the diastole. The fact that during exercise such as walking, the venous pressure does not drop or drops only insignificantly is to be attributed to this mechanism. The raised venous pressure leads to increased exudation of plasma into the extravasal space, to edema and to the other symptoms described above.

2. Increased strain caused by augmentation of blood flow results in dilation of the superficial veins and, eventually, loss of valves. The "private circulation" is less marked than with primary varicosis. It may be absent more centrally or entirely and the blood may flow centralward in the saphenous vein as long as the deep system is still occluded. What is decisive, however, is the factor of "reflux", which occurs after recannulation when the patient rises from lying to standing. Such reflux need not be confined to the superficial veins. Under conditions of extensive recannulation it may also occur via the deep veins.

These two pathogenetic mechanisms are reflected by different findings in *instrumental investigation.*

1. There is no evidence of pressure drop and decrease of the blood content in the tissue during motion. The normal drop of pressure by no less than 50% of the original level is not attained any more. In that case the muscular pump is impaired.

2. The pressure recovery time is significantly shortened, as is the refilling time of the tissue, which may be measured by light-reflection rheography or plethysmography. In addition thereto, Doppler-probe readings obtained while the patient sits up from a supine position indicate retrograde bloodflow. All these readings are evidence of *reflux*. Differential diagnosis may be established in the following manner:

Reflux disappears after application of a tourniquet or a sphygmomanometric cuff inflated to 50 to 80 mm Hg. In that case we are dealing with *isolated incompetence* of the *superficial veins* as observed under conditions of primary varicosis. If, after application of the sphygmomanometric cuff, reflux is still present distal to it, there is *incompetence of the deep veins,* for which evidence may also be obtained by means of a Doppler-probe.

Finally, *combined incompetence* of both superficial and deep veins may be found in severe cases.

Evidence of reflux may also be obtained by venography. By ascending or retrograde straining phlebography, the condition of valvular incompetence may be classified into various degrees (classification according to *Ferris* and *Kistner,* 1982).

Degree 0: no reflux. Degree 1: reflux in thigh only. Degree 2: reflux down to thigh and calf. Degree 3: reflux down to calf. Degree 4: massive reflux down to calf.

Apparently due to the strain exerted on the *perforating veins* as communicating pathways to the superficial veins and later also under conditions of secondary varicosis, the post-thrombotic syndrome often is combined with *dilation and incompetence of perforating veins.* While the role of perforating veins within the scope of primary varicose veins ist still disputed – the predominant direction of the bloodflow being from the surface to the deep system (*Bjordal,* 1972 b) –, incompetent perforators may significantly contribute to the development of leg ulcers under conditions of secondary varicosis. This is because they carry blood mainly from the deep system to the superficial system, thereby contributing to the increased pressure in the superficial skin veins of the supramalleolar region.

In the veins of the upper and lower leg the post-thrombotic syndrome is mainly caused by impairment of the muscle pump and by reflux. The situation is quite different in the state after *thrombosis of the pelvic veins.* In view of the fact that the majority of pelvic veins is valveless, reflux does not play a significant role. The state after isolated thrombosis of pelvic veins is instead characterized almost exclusively by *scarcity of vessels.* Pelvic veins must carry off considerably higher quantities of blood per minute than peripheral leg veins.

In normal condition their diameter is significantly higher than that of peripheral veins. The resistance in a vessel changes by the power of four of the radius. While in most cases of thrombosed or partially recannulated pelvic veins far-flung collateral circulation will form – and may be identified as *suprapubic collaterals* by inspection –, the diameter of a healthy iliac vein is never equalled. This is why the post-thrombotic condition in the pelvic area is characterized by a lack of vessels of an appropriate diameter.

The situation may be considerably complicated if there is post-thrombotic state of both *leg and pelvic veins.* Simultaneously, there may be impairment of the muscle pump, reflux and congestion caused by scarcity of appropriately large offcarrying vessels in the pelvic area. These supervening factors may even be aggravated by the role of perforators in the development of post-thrombotic ulcers. While precise differentiation between these individual components is of minor importance for measures of conservative treatment, it is highly mandatory for surgical therapy.

c) Therapy of the Post-thrombotic Syndrome

If carried out consistently, conservative therapy of the post-thrombotic syndrome may be remarkably successful. In view of the fact that the symptoms all represent consequences of venous congestion, treatment must aim at eliminating the congestion. This may be achieved in part by measures *Luke* (1951) summarized by the expression "new way of life", including:

1. Elastic compression of the extremity at daytime;

2. the patient should not stand for longer periods than 30 minutes at a time;

3. the patient's daily routine should be interrupted for half an hour twice or three times a day for the purpose of keeping his legs elevated;

4. when sitting, the patient should keep his legs elevated on a chair or a footrest;

5. the foot of the patient's bed should be elevated by about 15 cm;

6. the patient should apply inert cold cream to the skin of the affected leg approximately every other day;

7. he should avoid any irritation of the skin caused by sun, hot-water bottles etc.;

8. he should be careful, as far as possible, not to cause lesions to the affected leg by knocking against furniture etc.

As a rule, ulcers heal quickly under conditions of bedrest and elevation of the leg. The process of healing is supported by compresses and possibly by antibiotics. Skin grafts may shorten the process of healing but sometimes one is surprised to see how fast the ulcers heal even if the transplanted skin fails to take. The decisive therapeutic factor is that of bedrest. – When the patient is reambulated and returns to his work, the ulcers will recur quickly unless the rules of the "new way of life" are obeyed.

In recent years various procedures of surgical treatment have been recommended. These methods will be discussed in Part V of this book (pp. 183 ff.). Here, we would but anticipate that while by reconstruction of valves reflux may be diminished, it is much more difficult to eliminate the impairment of the muscle pump.

d) Raju's Concept

On the basis of extensive angiographic and instrumental examinations, *Raju* (1983) developed a concept, which substantially deviates from views generally held, as well as from the opinions set forth in this book.

Raju examined 147 patients suffering from venous complaints, including 14% of varicose vein patients and 14 healthy subjects without any history of venous diseases, by the following methods:

Doppler-ultrasound measurement with and without constriction of the superficial system; venous pressure measurement (method according to *Nachbur,* 1977) with and without constriction, before and after exercise; impedance and photoplethysmography; ascending and descending phlebography. Surprisingly, isolated incompetence of the superficial system was not found in a single case. Instead, isolated incompetence of the deep veins was revealed in 70 to 80%. A vast majority of these patients did *not* show any evidence of previous thrombophlebitis. – In some 25% of the patients examined, a combination of superficial and deep valvular incompetence was observed. Obstructions were found in no more than 13 out of 147 patients. There was evidence of reflux, however, in as many as 120 cases. – Among the 14 apparently healthy subjects, two had reflux in the deep venous system and 12 were free from reflux. – The locations of reflux may be seen from the following table.

Table 5. *Anatomic level of reflux in 100 patients*

	Left side	Right side
Femoral vein	97%	98%
Popliteal vein	75%	79%

From these findings *Raju* draws the following conclusions:

He rejects the traditional views of both the Theory of Private Circulation according to *Trendelenburg* and the Theory of Perforators according to *Linton* (1938), and *Cockett* and *Dodd* (1976). He furthermore rejects the classification of valvular incompetence into that of the superficial and that of the deep venous system. Valvular ectasia is regarded mainly a consequence of a congenital defect rather than a result of thrombosis. Thrombosis of deep veins is not the cause but the consequence of a primary valvular defect. *Raju* postulates the following sequence of events in a chain reaction:

reflux,
stasis,
phlebitis, and
destruction of valve.

In this sequence of events functional impairment of the valve increases reflux, whereby the circle is closed and the chain reaction recommences.

The following arguments may be held against this concept:

1. A certain selection may have unintentionally occurred when *Raju's* patients were collected (the low number of cases with clinical symptoms of primary varicosis would be highly indicative of such selection). *Raju* himself admits that possibility in the discussion.

2. In our view, the possibility of isolated valve-pocket thrombosis is not adequately allowed for by *Raju*. Thrombosis of valvular sinuses may remain unnoticed in patients surviving the illness, but are often detected postmortally.

3. According to several authors (see pp. 75, 162), reflux frequently may occur without causing the slightest symptoms. The 2 cases of reflux revealed among 14 symptom-free subjects, as reported by *Raju,* are telltale evidence of this.

4. The differentiation between superficial reflux and functional impairment of deep veins is based on the direction of flow in the perforating veins and in the saphenous system. This differentiation has proved its value for therapeutic purposes. Therefore, they should not be abandoned unless *Raju's* findings are confirmed by other investigators, on the basis of non-selected patient material.

Summary – Part III

Pathologic processes in venous valves may be efficiently examined by the method of en-face preparation. In silver-stained specimens, damages of the valves are of different appearance than damages of the venous wall, due to the fact that the valves are devoid of smooth muscles. Mechanical and chemical lesions are discussed. In studies of age-caused changes it is conspicuous that the valvular endothelium is less involved than the endothelia of the venous wall. Sinuses of venous valves are often seats and points of origin of thrombosis. Post-thrombotic changes were studied in en-face preparations. They are largely similar to changes found in the uppermost valves of the saphenous veins under conditions of varicosis. Functionally, there are fundamental differences between primary and secondary varicose veins. The "Theory of Private Circulation" and the "Theory of Perforators" are critically compared to each other as explanations of the development and basic concepts for the treatment of varicose veins. – The concept of Raju, which conflicts with generally held views on the development of valvular incompetence of superficial and deep veins, is discussed.

Part IV. Radiology of Venous Valves

A. Methods and Results

In recent years various methods of correcting incompetent venous valves by surgical therapy have been developed. For such purposes direct radiologic visualization of venous valves is required. By *ascending phlebography,* the valvular sinus is clearly visualized. The diagnostic accuracy is enhanced by *retrograde straining phlebography.*

In former years we recommended puncturing the femoral vein in the inguinal region while the patient is in a standing position, and telling the patient to strain during the injection (according to *Martinet,* 1959). If the valve is competent, the contrast medium is prevented from proceeding further in distal direction. This is beyond doubt the most accurate method. Unfortunately, patients standing upright and requested to strain, often collapse while the contrast medium is injected into the groin. For this reason we abandoned the method completely. What we do require from the radiologist in most radiologic examinations of the deep veins is accurate visualization of the valvular status. As a routine method which is up to that standard we use a much simpler procedure which was also described by *Hach* (1981). Normal ascending phlebography, application of a tourniquet above the ankle, injection of contrast medium into one of the veins on the dorsum of the foot. The patient is in a recumbent position of 65°. The flow of the contrast medium is monitored on the television screen. When the contrast medium has reached the popliteal vein, we take the usual two a.p. and lateral routine shots. Thereafter, we request the patient to strain forcibly: while the patient is doing so, the valves of the lower leg and calf veins are accurately visualized. The procedure is repeated when the column of contrast medium has reached the groin. Thus, not only the terminal valve of the long saphenous vein, which is of utmost functional importance, but also the valves of the femoral vein are well visualized (Figs. 79–81).

The competence of valves in deep veins, in particular in the femoral vein, is visualized with particular accuracy by the method of "Phlebographic *Trendelenburg's* test" according to *Gullmo.* While the

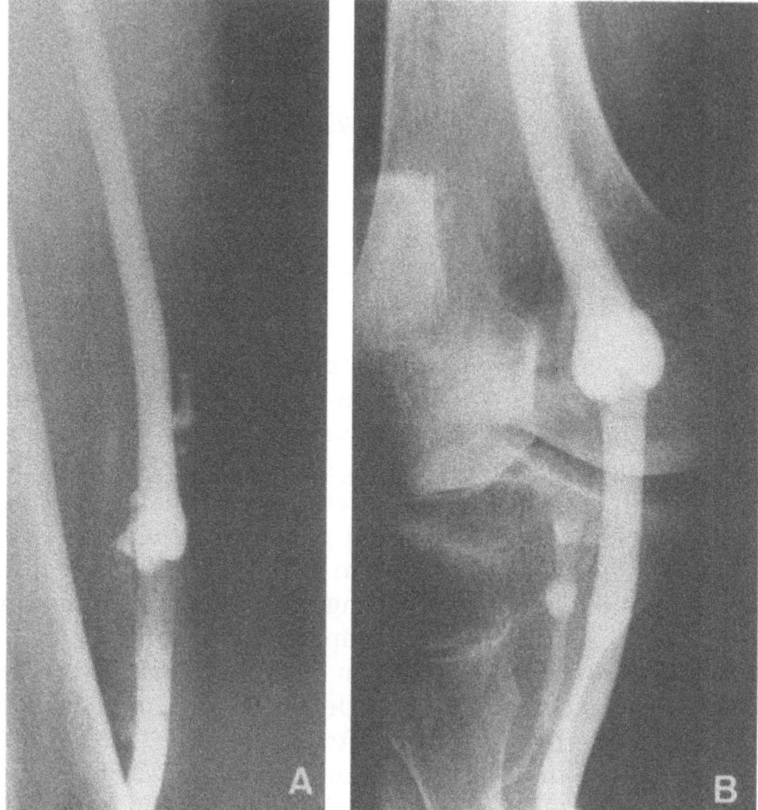

Fig. 79. **A** Competent valve in femoral vein. Ascending straining phlebography. **B** Competent valve in popliteal vein. Visualized by ascending phlebography without straining

patient is in a supine position, the femoral vein is punctured in the groin. During the injection, the patient strains, while we take several shots. *Gullmo* (1964) emphasizes that the contrast medium even passes through fully competent valves when *Vasalva's* maneuver is repeated in quick succession. *A single and prolonged Vasalva's maneuver produces a total stop by competent valves* (see p. 76, Fig. 39, "Function of normal venous valves").

We recommend this technique as a method of control before and after venous valve surgery.

Valves in superficial veins are well visualized by direct puncturing of these veins in the peripheral region while the patient is in a slightly reclined position. This examination is of importance if the vein is to be used for bypass surgery. In that case we would urgently

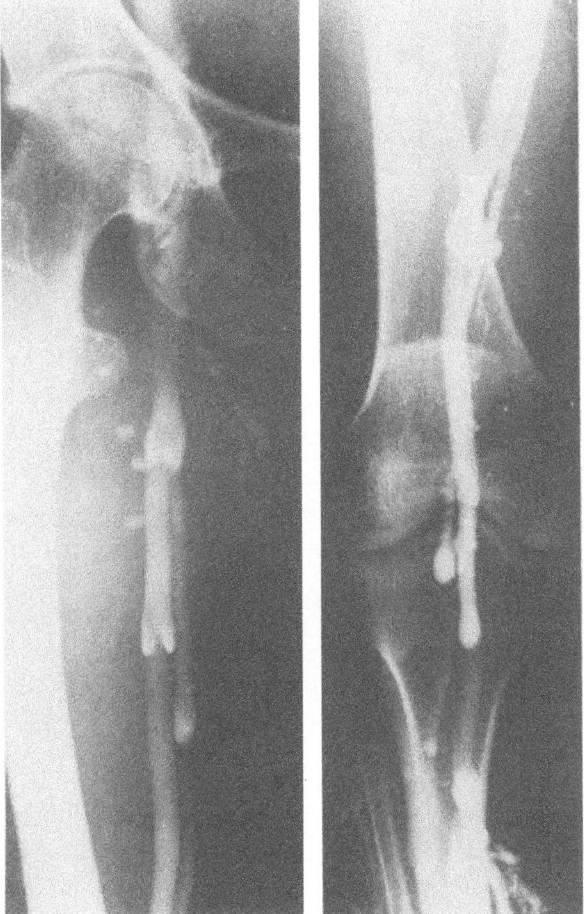

Fig. 80. Competent valves in femoral and popliteal veins. Visualized by ascending straining-type phlebography

recommend using one of the newly developed low-osmolal contrast media in order to preserve an undamaged intima.

Visualization of Perforating Veins

Most authors recommend visualizing perforating veins by filling of the deep veins. This is certainly correct from a physiological point of view. If the valves are competent, the backflow of the contrast medium to the surface is frequently stopped by the valve that is located near the passage through the fascia.

Nevertheless, we believe that the procedure recommended by us (*May* and *Nissel*, 1973) is clearly preferable for practical use.

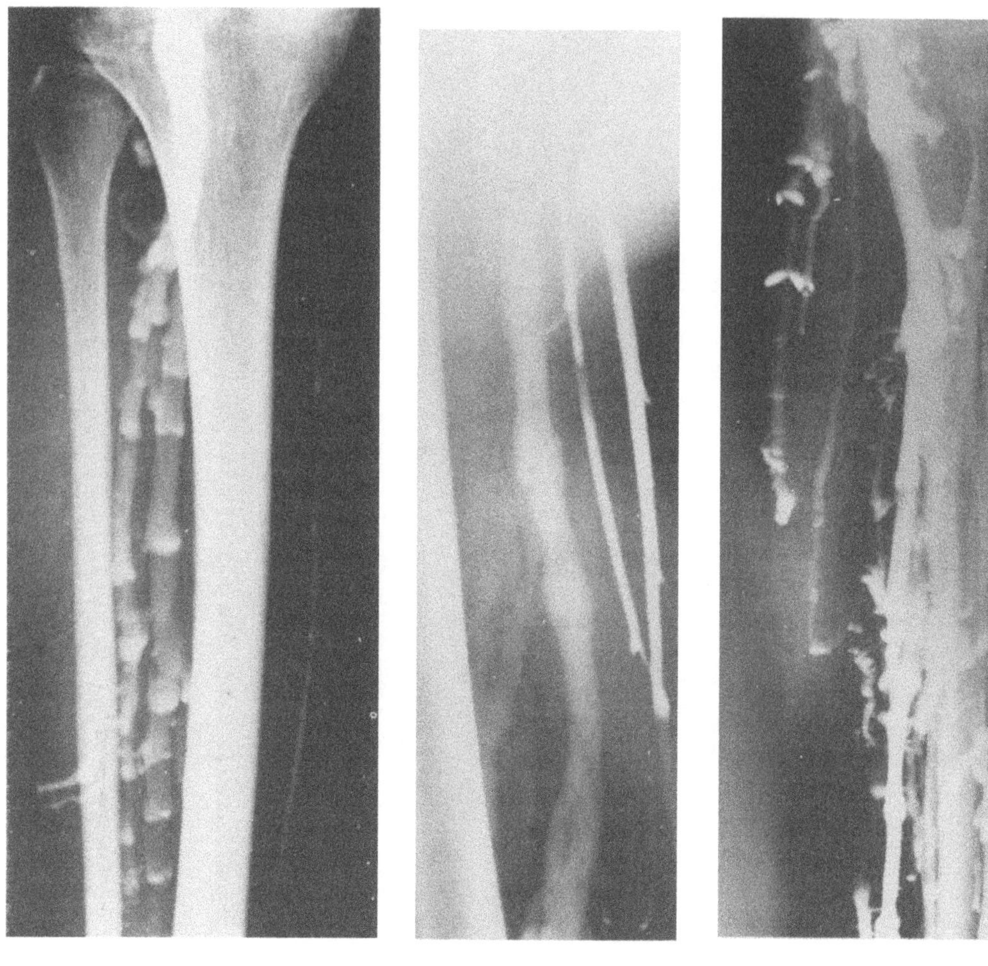

Fig. 81   Fig. 82   Fig. 83

Fig. 81. Valves in lower leg veins. Ascending straining-type phlebography
Fig. 82. Competent superficial veins
Fig. 83. Straining-type phlebography. Valves in gastrochemic veins

While the patient is in a supine position, the long saphenous vein is punctured approximately at ankle level. The foot-end of the table is slightly elevated. After a tourniquet has been applied approximately two hands' breadths above the site of injection, contrast medium is injected. After visualization of the first segment, the tourniquet is moved in a proximal direction step by step. Thus, we may visualize all perforating veins almost up to the groin in a comparatively reliable manner. The system of the short saphenous vein must be visualized separately, of course. Thread-sized perforating veins are considered to be practically competent. Usually, however, the valves are visualized directly.

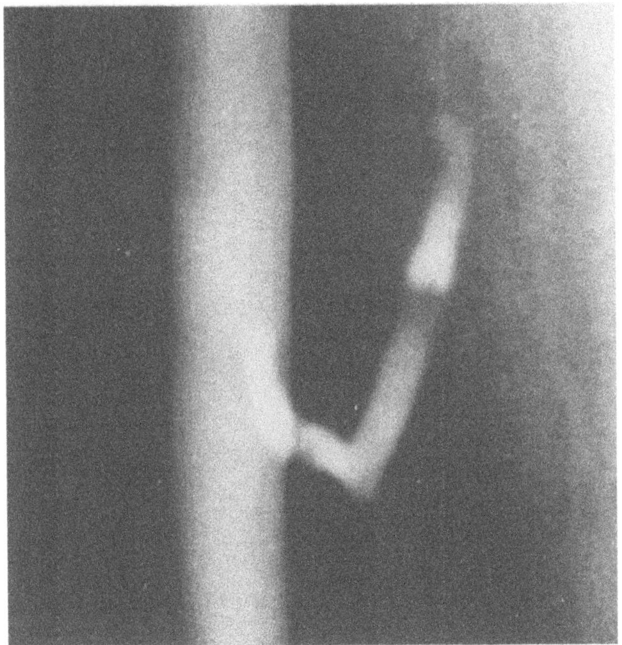

Fig. 84. Valves in perforating vein of thigh

B. Lesions Caused by Contrast Medium

a) Pathogenesis

If injecting conventional contrast media into varicose superficial veins, one may often notice that as soon after as the very next day the vein will show all symptoms of phlebitis (*May* and *Mignon,* 1978). – In varicose veins this is of minor consequence and occasionally even saves the trouble of sclerosing therapy. We suspect, however, that similar effects might occur as a consequence of filling deep leg veins by contrast medium. In particular, the region of the valvular sinus is endangered.

Radiopaque contrast media have higher specific weights than blood, due to their content of iodine. If the injection is made while the patient is in upright or reclined position, sedimentation effects occur. The contrast medium shows a tendency to collect at the lowest point. Such sedimentation occurs even against the current of blood.

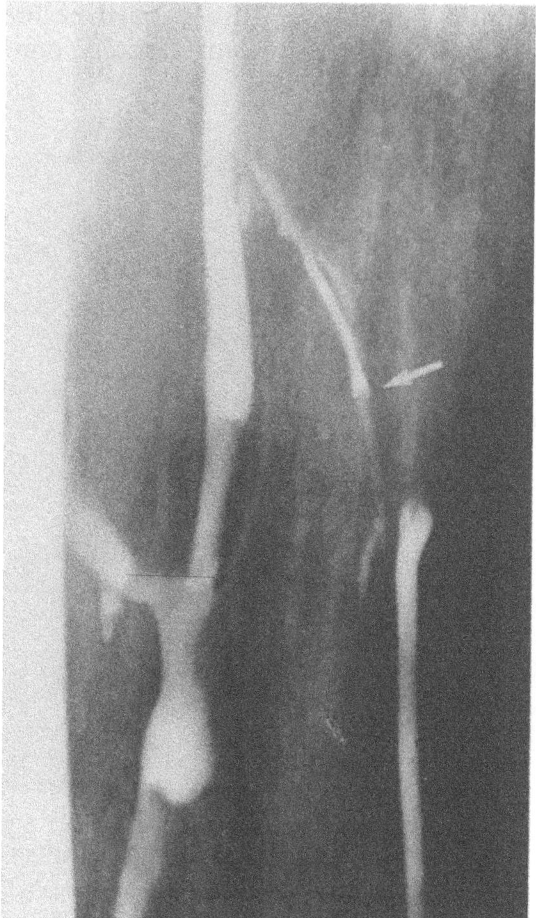

Fig. 85. Ascending straining-type phlebography. Competent valve in perforating vein, connecting long saphenous vein and femoral vein, visualized excellently

Thanks to these physical properties of radiopaque contrast media, the region of the valvular sinus is "stained" with particular intensity. The valvular sinus is the lowest point of a venous segment. There is almost no current of blood in the valve pocket, irrespective of whether the valve is open or closed. Due to these circumstances, the region of the valve is often visualized with enhanced contrast, and the staining of the valvular sinus may be maintained while the contrast medium has been flushed out of the other regions of the vein by the blood following behind.

The valvular sinus is the locality of minimal resistance in the vein for two reasons:

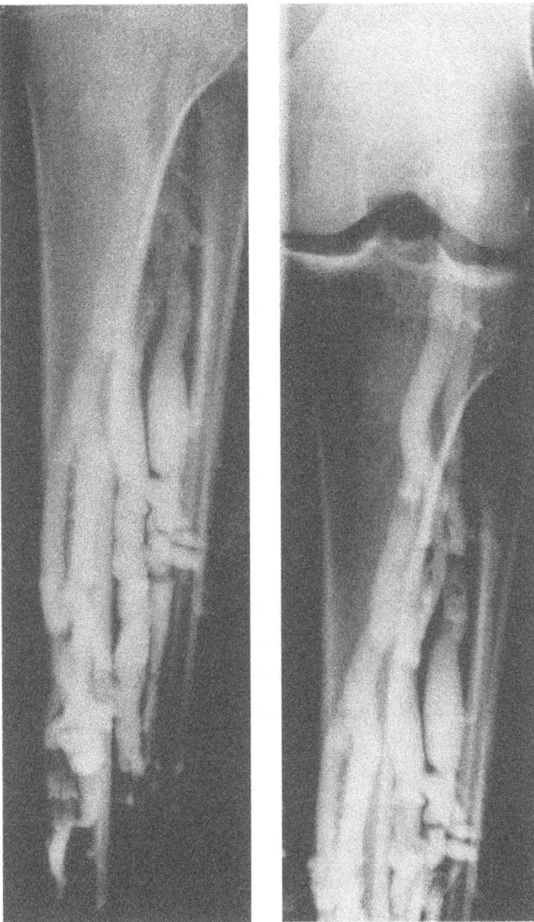

Fig. 86. Valvular incompetence in lower leg veins of patient overweight by 30 kg

1. The contrast medium sedimentates in this area and is present in high concentrations;

2. there is almost no current of blood, which offers excellent conditions for the development of local thrombi.

Due to the fact that lesions caused by contrast medium predominantly occur in the region of the valvular sinus, they evade clinical diagnosis. Initially, development of thrombi in the valve pockets of peripheral veins will hardly manifest itself clinically. The hemodynamic effects caused by the loss of the valve will be of minor significance as long as there are still healthy valves in more proximal locations. It must be emphasized that even the post-thrombotic syndrome does not occur immediately after the end of a thrombotic process but may take years to become manifest. Therefore, when visualizing valves by peripheral phlebography, we must be aware that we may cause a considerable percentage of undetected valve lesions.

b) The Role of the Osmotic Pressure

In phlebography side effects consisting in thromboses of deep veins or inflammation of superficial veins are exclusively due to the high osmotic pressure of the various contrast media. Other causes of side effects such as *lipophilia* or the protein binding capacity of the contrast media are of minor importance for phlebography. We therefore will not deal with these questions here, nor with the question of allergy or hypersensitivity to contrast media.

We would stress in this respect that, in our view, the damages caused by the osmotic pressure of conventional (ionogenic) contrast media are often underestimated so that these contrast media are applied in harmful concentrations. This view is supported by the following statements found in specialized literature on the subject.

Albrechtsson and *Olsson* (1976) tested for post-phlebographic thromboses by the iodine-fibrinogen test and obtained positive findings in an alarming number of cases – even if subjective complaints were not stated.

In the patient material investigated by *Lea Thomas* in 1976, thromboses were found in *only* 0.54% of cases. In 1978, however, *Lea Thomas* and *MacDonald* reported that in 25% of the cases examined there was pain in the injected extremity indicative of thrombosis. In a further 5% of the cases thoracic pain – indicative of pulmonary embolism – was stated. *Cranley* (1975, quoted according to *Laerum,* 1983) reported thrombophlebitis in 50% of cases. – *Bettmann* and *Paulin* (1977) found clinical evidence of phlebitis in 32% of cases after phlebography, in which 60% diatrizoate was used. By reduction of the concentration to 45%, the percentage of thrombosis decreased to 10. In 1983, *Laerum* and *Holm* carried out a double-blind test. Two contrast media with concentrations of 280 mg iodine per ml (approximately 58% solutions) were tested. One was a calcium-methylglucamin-metrizoate (Isopaque cerebral), the other the low-osmolal metrizamide (Amipaque). After phlebography ^{125}I-fibrinogen uptake tests were made. All patients with suspected thrombosis were phlebographed a second time. In those patients who had received the high-osmolal contrast medium, the fibrinogen test was positive in 43%. Groups of 5% each were not further examined or showed other reasons for the result of the test. Of the remaining 33% of positive cases (14 patients), 11 showed definite phlebographic evidence of thrombosis, i. e. 26% of the entire patient material. After the use of metrizamide, merely 2% of the patients who had received that substance showed definite evidence of thrombosis in fibrinogen test and phlebography. The difference between the two contrast media was highly significant. Significant differences between high- and low-osmolal contrast media were also found when the local pain, sensitized during injection was compared.

LePage (1985) described three cases which involved slough of major areas of the dorsum of the feet ranging in size from 20 to 35 cm². One additional case even developed gangrene and necrosis requiring amputation of the forefoot. These lesions were produced with undiluted 76 or 60% contrast media, undoubtedly concentrations that should be avoided in peripheral phlebographies. Extravasation may play a major part in causing such injuries.

In addition to clinical experience, experimental results are also clearly indicative of the risks of using high concentrations of contrast medium as well as of the enhanced tolerance of low-osmolal contrast medium. First reports according to which contrast media and other high-osmolal solutions may cause lesions of endothelium were published by *Zinner* and *Gottlob* (1959) and *Gottlob* and *Zinner* (1964). By using the method of en-face preparations and endothelial silver staining, the difference between high-osmolal and low-osmolal contrast media was demonstrated.

An interesting method of examining the endothelial tolerance of radiopaque contrast media is a test developed by *Laerum et al.* (1983) during which cell cultures of umbilical cord endothelia are subjected to the solutions. This test, as well as several other methods which cannot be described in detail here, clearly demonstrated the superiority of low-osmolal contrast media.

T. *Almén* deserves credit for having introduced low-osmolal non-dissociating radiologic contrast media available to us.

c) Low-osmolal Contrast Media

Like conventional dissociating (ionogenic) contrast media, non-dissociating contrast media consist of a tri-iodized benzene-ring to which various lateral chains are substituted. Low-osmolality is brought about by the fact that the side-chains are bonded firmly to the benzene-ring so that no salts that might dissociate in watery solutions are formed. In this way, the number of dissolved particles is considerably reduced so that the osmotic pressure is decisively lower.

The first non-ionogenic contrast medium was Metrizamide (Fig. 87 B), whose endothelial tolerance was in sharp contrast to

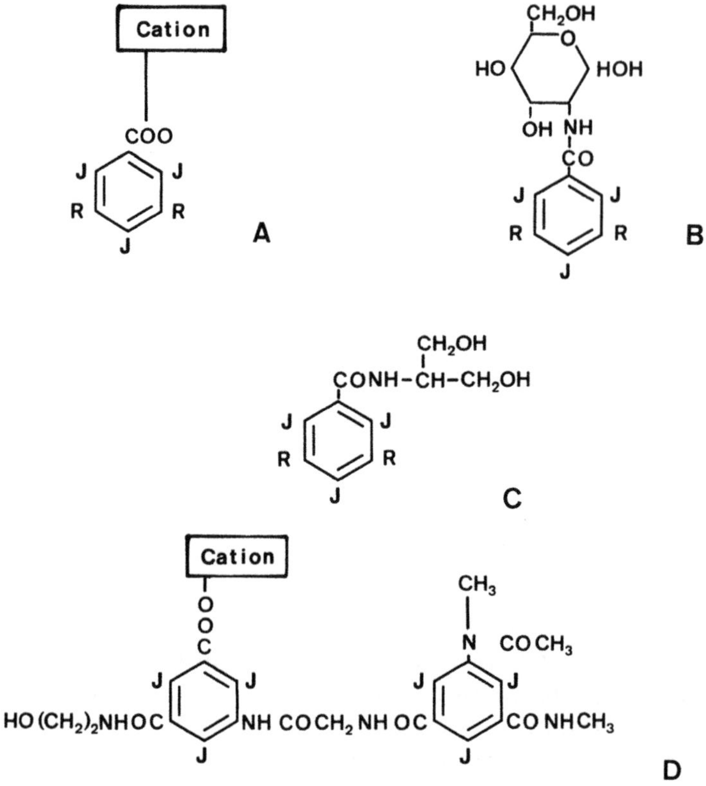

Fig. 87. Various types of radiologic contrast media for angiography. **A** Conventional contrast medium such as Angiografin or Conray. **B** Amipaque, the first non-dissociating contrast medium. **C** Iopamidol and **D** Hexabrix, a dimer – only one component of which is salt-forming. The osmotic properties of B, C and D are roughly equal

relatively poor solubility and an excessively high price. More recently developed monomeric low-osmolal contrast-media are Iopamidol and Iohexol. Iohexol was also introduced by *Almén*. In addition thereto a dimer (Ioxaglate) is available, which consists of two tri-iodized benzene-rings, only one of which forms salt. Therefore, by dissolution of one molecule-containing 6 iodine atoms two dissociated particles are generated, which is a ratio equalling that found in the monomerics mentioned above (Fig. 87 D). – Iopamidol, Iohexol and Ioxaglate, a dimeric, in contrast to Metrizamide offered the advantage of better solubility in water; they are available in ready-for-injection ampoules. Besides, their price is less prohibitive than that of Metrizamide. Table 6 gives a survey of the generic and commercial names, the osmotic pressures and the prices of these contrast-media.

Table 6. *Contrast media (CM) currently in use*

Type of CM	Generic Name	Commercial Name	Osmotic Pressure (In mosm/Kg at 300 mg Iodine/ml)	Price* (Average of conventional ionogenic CM = 100)
Conventional = Ionogenic = Dissociating CM	Amidotrizoate (Diatrizoate)	Angiografin Hypaque Renografin Urografin		
Monomers	Ioglicate Iothalamate Ioxitalamate	Rayvist Conray Telebrix Vasobrix	1500–1600	100
	Metrizoate Iodamide	Isopaque Uromiro		
Dissociating Dimer	Ioxaglate	Hexabrix	560	218
Non-ionogenic Monomers	Metrizamide	Amipaque	480	8603
	Iopamidol	Iopamiro Solutrast	616	503
	Iohexol	Omnipaque	690	458
	Iopromide	Commercially not yet available		
For comparison: Blood			290	

* 1985 in Austria.

Comparable to Iopamidol is Iopromide, a contrast-medium which will be marketed in the near future.

As may be seen from Table 6, the osmotic pressure of the new low-osmolal contrast media is still twice that of blood. Excellent endothelial tolerance was proved by *Nyman* and *Almén* (1980) and by *Gottlob* (1978 b, 1979).

d) Prevention of Lesions Caused by Contrast Media

Primarily, contrast-medium lesions may be prevented by the use of low-osmolal contrast-media. This principle has its limit in the higher price of such contrast-media, however. On the other hand, conventional contrast-media may be used without risk if high concentrations are avoided. We believe that 60 to 80% radiologic contrast-media are counterindicated for the visualization of peripheral veins and venous valves. As a rule, substantially lower concentrations of 30 to a maximum of 40% will suffice for such purposes. Another drawback of high concentrations is that fine details, e.g. of the valvular apparatus, will be obscured and hidden from closer examination, whereas they may be well visualized by means of low concentrations.

Alison et al. (1985) found that for venographic examinations of peripheral leg veins a 20% ioxaglate is sufficient. If also the pelvic veins are to be visualized, a 32% concentration of the contrast-medium is sufficient.

Our recommendation: One should always try to make do with a 40% concentration of contrast-medium. Another important aspect is that higher concentrations of solutions should not be diluted by saline but distilled water, which is the only way of optimally reducing osmotic pressure.

Another method of reducing the dosage of contrast-media is injecting a low volume. It is not always necessary to aim at visualization of the entire venous system of the leg. If no more than information on the condition of the valvular apparatus of a given region is required, it sometimes will suffice to inject a small volume of contrast-medium near the region to be visualized. The decisive question here is what information does the clinician require for diagnosis or therapeutic purposes?

In addition thereto, the risks of phlebography may be kept down by seeing to it that the injected contrast-medium is flushed out of the veins as quickly as possible. After application of contrast-medium, some authors inject saline with or without heparin. The best method of flushing contrast-medium out of veins, however, is encouragement of bloodflow. Such encouragement of bloodflow is achieved most easily by having the patient perform physical exercise after the examination; such exercise should consist of movements enhancing the bloodflow in such a manner that the contrast-medium will disappear quickly, e.g. several toestands in rapid succession or walking immediately after application of contrast-medium. Another safety factor may be seen in the application of elastic bandages.

The question whether the prevention of post-phlebographic lesions should also include the administration of anticoagulant drugs is still contested. In a double-blind experiment *Laerum* (1981) found unequivocal advantages.

Allergic reactions: Although such accidents are very infrequent one should always be prepared for their occurrence and keep ready everything necessary for first aid, including artificial respiration, access to the veins even if they are collapsed (venesectio), and a supply of intravenous injectable cortison derivatives and drugs stimulating circulation.

Summary – Part IV

For visualization of the valves in leg veins, ascending and retrograde filling of contrast medium may be used. Straining-type phlebography according to Gullmo is recommended in view of the fact that the tendency to collapse is eliminated when the patient is lying. However, injection of contrast-medium into standing patients or patients resting in semireclined position offers the advantage of dilation of the valvular sinus. Competent valves completely stop the flow of contrast medium in most instances. Accurate visualization is important for contemplated surgical reconstruction of valves.

Conventional contrast-media are hypertonic solutions causing endothelial lesion. Due to the fact that there is almost no current of blood in it, the valvular sinus is the location of minimal resistance, where radiologic contrast medium settles down by sedimentation and may stay for several minutes. For the prevention of presumably frequently occurring but rarely detected damages to the valvular apparatus, it is recommended to use low-osmolal contrast media, avoid higher concentrations of contrast media and make sure that the contrast medium is flushed out of the veins as soon as possible.

Part V. Importance of Venous Valves in Surgery

1

Venous Valves in Arterialized Veins

A. Autologous Veins

a) General

Transplantation of veins into the arterial system in the form of bypass-surgery has become a routine operation in many parts of the body. The most frequent applications are aortocoronary bypass performed in order to shunt an obliterated segment of the coronary arteries, and bypass of peripheral leg arteries to shunt an obliterated segment, mainly in the region of the femoral artery.

Venous valves would constitute considerable obstacles in a bypass of the above nature. Obstruction of the bloodflow by venous valves may be prevented by two methods:

1. In the manner first suggested by *Kunlin,* the vein is reversed so that the venous valves do not form an obstacle to the bloodflow any longer.

2. It is possible to destroy venous valves mechanically.

b) Function of Venous Valves in Reversed Veins

Venous valves can be grafted into the arterial system in such a direction that the valve is kept open by the bloodflow. In this case they not only let the bloodflow unhindered but occasionally may even promote bloodstream. They do so by preventing reflux, which in some vessels occurs under physiologic conditions. Such reflux is found in the coronary arteries of the left ventricle during the systole. It is caused by the fact that, during the systole, the smaller arteries are compressed, whereby a short backward current develops in the larger coronary vessels. In peripheral leg arteries there is reflection of the pulse waves at the end of the systole and at the beginning of the diastole, whereby a short wave in centralward direction is caused. In the presence of a functional valve, the reflux may be prevented and the flow per minute through the respective vessel increased. In peripheral leg arteries reflux develops, in particular if the reflection of the pulse waves is strengthened by stenosis. The usefulness of venous valves for the prevention of such reflux was demonstrated by *Kantrowitz* and *Lerrick* (1952). *Weissenhofer et al.* (1974) proved the valuableness of functional venous valves in peripheral leg arteries by flowmeter measurements, and *Baba et al.* (1976) furnished experimental evidence of similar valuableness in the aortocoronary bypass.

Weissenhofer et al. pointed out, however, that the useful effect was provable for a short period after the operation only and that the valves became dysfunctional within 3 to 4 weeks after the operation. The valves were histologically degenerated, they had shrunk and thickened so that they had become incapable of being functional.

We must ask ourselves whether such degeneration of venous valves is inevitable (cf. endothelium-preserving operational method, pp. 188 ff.).

c) Valve-caused Stenosis in Arterialized Veins

We have already mentioned above that the valvular pocket is the site of minimal resistance in the entire transplanted vein. This is due to the fact that there the bloodflow is slowest and occasionally it has no flow at all. No wonder then, that in transplanted veins not only pocket thrombi were found (*Harjola et al.*, 1982) but also massive fibroses in the regions of the venous valves. Occasionally, the grafted vein was even observed to have become completely obliterated (*Breslau* and *De Weese*, 1965; *Downs*, 1971; *McNamara et al.*, 1967; *Szilagyi et al.*, 1973; *Welsh et al.*, 1974). *Breslau* and *De Weese* were the first to excise such a fibrotic nodule from the grafted vein, and succeeded in restoring patency of the vessel.

d) Elimination of Venous Valves for *in situ* Bypass Surgery

It is one of the drawbacks of reversed-vein bypass that the lumen of the graft is wider at the distal end than at the proximal end. If the vein is left in situ, the lumen of the vein decreases toward the periphery in a similar manner as the natural lumen of arteries. In that case, however, it is necessary to destroy the valves after ligation of the venous branches. This may be attempted by insertion of blunt probes, e.g. by insertion of a stripper as recommended by *Rob* (quoted according to *Hall*, 1962). *Hall* himself, the "inventor" of in-situ bypass, performed transverse incisions whereever a valve was impeding the bloodflow and removed the valvular cusps at these sites by direct excision. *Conolly et al.* (1964) invaded the vein from its distal end by a stripper and pushed the instrument forward to the distal end three times, whereupon they inserted an endarterectomy stripper of appropriate size into the vein in the same direction. *Samuels et al.* (1968), on the other hand, disapproved of the use of strippers due to the fact that often the vein is damaged by such instruments in the regions of entries of tributaries and in one case the intima had also been torn up over a distance of several centimeters. They designed a special instrument to be inserted in proximal direction, consisting of medium-gauge wire, the upper end of which was provided with two cylinders. At the distal end of the upper cylinder there were two blunt hooks. They were arranged in such a fashion that, when the wire was pulled back, they would grasp into the valvular pockets and, in this way, destroy the cusps. The two cylinders served the purposes of straightening the instrument and dilating the vein.

Samuels et al. stated in their paper: "It is noteworthy that the closing lines of the valve cusps are orientated parallel to the skin surface" – and thus confirmed the findings of *Edwards* (1936), obviously without knowing his work (see p. 31 f.).

For venous valvulotomy *Langeron et al.* (1978) used a "bougie parapluie" as first described by P. *Cartier*. Several such "umbrellas" of increasing size were repeatedly pulled through the vessel in peripheral direction. The authors emphasized the risk of damaging collateral vessels.

Mills and *Ochsner* (1976) and *Leather et al.* (1981) used a miniaturized valvulotom, such as used in cardiac valve surgery. It is a 32 cm long wire, the top of which is bent to form a rectangular deflection of 3 mm length. The end of the deflection is thickend in a ball-shaped manner. The deflection is provided with a cutting edge, extending from the longitudinal portion of the wire to the ball. *Leather et al.* emphasize that – irrespective of the method employed – for purposes of valvulotomies the region of the valve to be destroyed must be exposed so that one can perform this work under visual control.

Corson et al. (1984) used a similar valvulotom and described the technique of the procedure in detail.

B. Venous Valves in Homologous Grafts

For purposes of comparison, *Gottlob et al.* (1976a) transplanted both autologous and homologous venous grafts into carotids of dogs. At

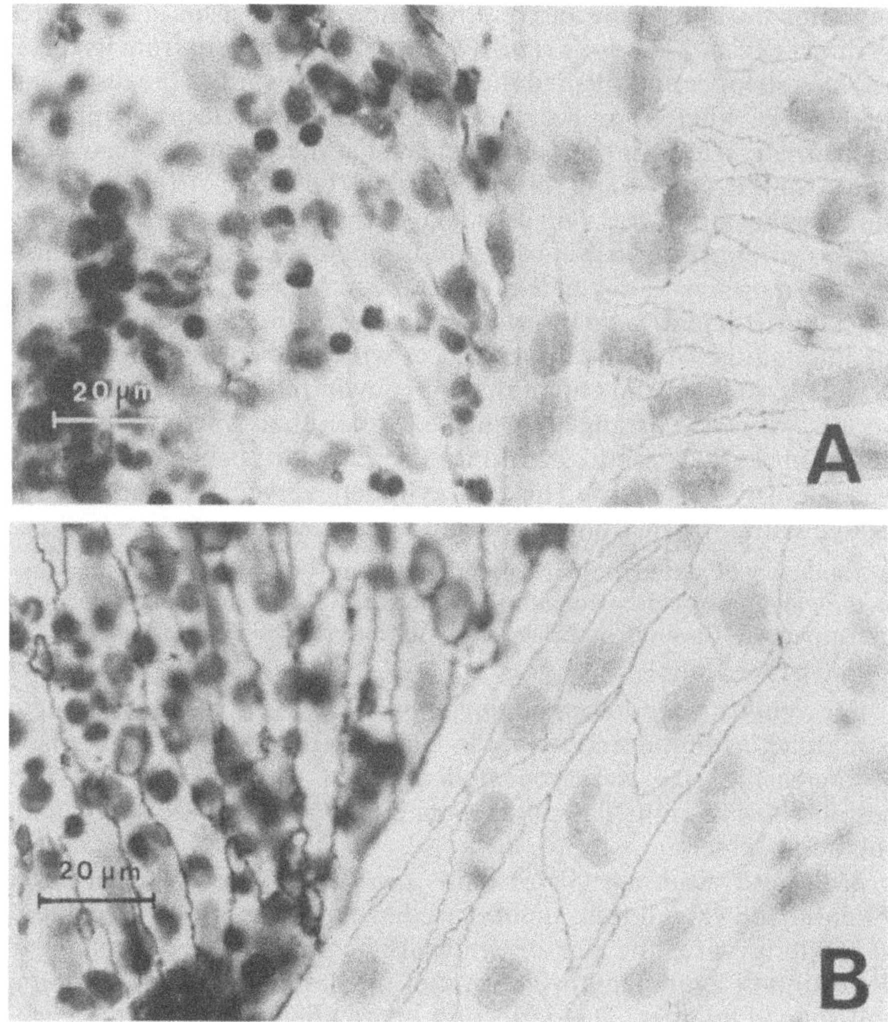

Fig. 88. Arterialized veins, homologous grafts. Valvular cusps on the right side, venous wall on the left side. Note that only the venous wall has been infiltrated by leucocytes

some time around the beginning of the second week, signs of rejection were observed in homologous grafts. There was occurrence of macroscopically visible parietal thrombotic plaques, which at the same time were no longer observed in autologous grafts of vessels. The regions of the anastomoses were invariably found to have been spared from rejection. It was assumed that by that time endothelium from the host vessel had already grown in. – Histologically, at sites without thrombi, the rejection was marked by infiltrations of leucocytes, which were found in the vascular wall below the endothelial layer. Minor infiltrations were also observed after autologous transplantations, but the homologous grafts were infested by leucocytes to far greater extent, the majority of the leucocytes being represented by round cells. What was conspicuous about these findings was the fact that in the regions of the venous valves no infiltrates whatsoever were found (Fig. 88). This was explained by the assumption that the infiltrates did not invade the subendothelial layer from the main current of the bloodflow but obviously via the vasa vasorum, which found connection to the host vessels as early as 2 to 3 days after the transplantation. The valves, however, not being in connection with the vasa vasorum, were spared.

2
Venous Thrombectomy

In arterial thrombectomy, a powerful current of blood indicates that all obstacles in the flowpath have been removed successfully. If, however, venous thrombectomy is performed, backflow from more proximal segments occurs only in the absence of competent venous valves at sites nearer to the heart. According to *Basmajian* (1952), valves are absent above the sapheno-femoral junction in 20% of the veins. According to *Di Dio* (1949), a valve is found in the common iliac vein in 7%; but most of these valves are small and incompetent. In the external iliac vein, valves are found in approximately 25%. According to *Basmajian,* more than ⅔ of these valves are competent.

Diethrich (1967) points out the complication which arisis if, after venous thrombectomy, there is no backflow of blood. For such cases *Diethrich* recommends inserting a clamp into the venous lumen in cranial direction and spreading it. Provided the vein is free from further thrombi, distinct backflow will, thus, be achieved after that. Instead of the clamp, a plastic catheter with wide lumen or the glass tube of an aspirator, disconnected from the appliance, may be inserted.

Haller (1961) emphasized the importance of valves for surgery of acute ileo-femoral thrombi. In 5 out of 24 cases he observed large valves in the external iliac vein.

3
Reconstruction of Venous Valves

A. General

As has been mentioned in the chapter describing the post-thrombotic syndrome, incompetence of the venous valves in the femoral and popliteal veins is the cause of congestion and edema, resulting in dermatitis, induration and ulcers in the ankle region and other portions of the lower leg. Incompetence of the valves of calf veins may lead to "failure of the peroneal muscle pump", which is of considerable importance as a causal factor of the complex of symptoms, comprised by the term "post-thrombotic syndrome".

At present, intervention by measures of surgery is largely restricted to attempts to reconstruct the uppermost valve of the superficial femoral vein or replace its function.

According to *Sandmann* (1982), the selection of therapeutic measures of surgery depends on the following typical patterns:

Type A: Excessive length of cusp edges due to loss of elasticity. When the venous pressure increases proximally, the valves yield, and there is reflux.

To this we would critically annotate that elastic elements are definitely of minor significance in valvular cusps, as has been demonstated by our electron-optical investigations. The changes referred to, seem to be attributable instead to degeneration of the collagenous framework.

Type B: Expansion of the valvular ring due to overdilation of the venous tube.

To these types we would add

Type C: Loss of valves due to previous thrombosis of the valvular sinus.

B. Plastic Procedures Apart from Transplantation or Transposition of Venous Segments

a) "Substitute Valve" According to Psathakis

This operational method consists of entwining the tendon of the gracilis muscle between popliteal artery and vein in such a manner that the vein will be constricted during contracture of the muscle. – This method did not meet with general acceptance.

b) Cuff-Method According to Hallberg (1972)

Knowing that valvular incompetence may be caused by dilation of the valvular ring, *Hallberg* sheathed the region of the incompetent valve by means of a plastic tube. While having achieved encouraging first results, *Hallberg* stresses that this method cannot be recommended for routine yet.

c) Valvuloplasty by Invagination of the Venous Wall

Eisemann and *Malette* (1953) produced valve-like structures in larger-sized veins by gathering folds at two sites of the venous wall opposite each other. The venous wall was gathered over a length corresponding to the threefold diameter, and the transverse fold created in that manner was tucked into the lumen.

d) Valvuloplasty According to Kistner

Kistner (1975) observed two forms of venous incompetence: cases with recannulation of deep leg vein thrombosis and extensive loss of valves, and cases with well-preserved valves, the incompetence of which was caused by elongation of the cusps. For the latter type of cases he recommended an operation for shortening the valvular cusps.

This method is illustrated in Fig. 89. Fig. 89 A is a schematic representation of a vein cut open in the region of a commissure. The cusps are too slack. B: an atraumatic 7 to 9–0 suture is stitched in from the outer side of the vessel on the level of the commissure. C: the suture is stitched in through the edge of the valvular cusp at a distance of 1 to 2 mm from the commissure. D: near the commissure, the suture is led back to the outer side of the wall. E: the suture is tied up at the outer side of the venous wall. F: a similar suture is made from the back side of the vein in the region of the commissure. G: this suture seizes those portions of both cusps which are near the

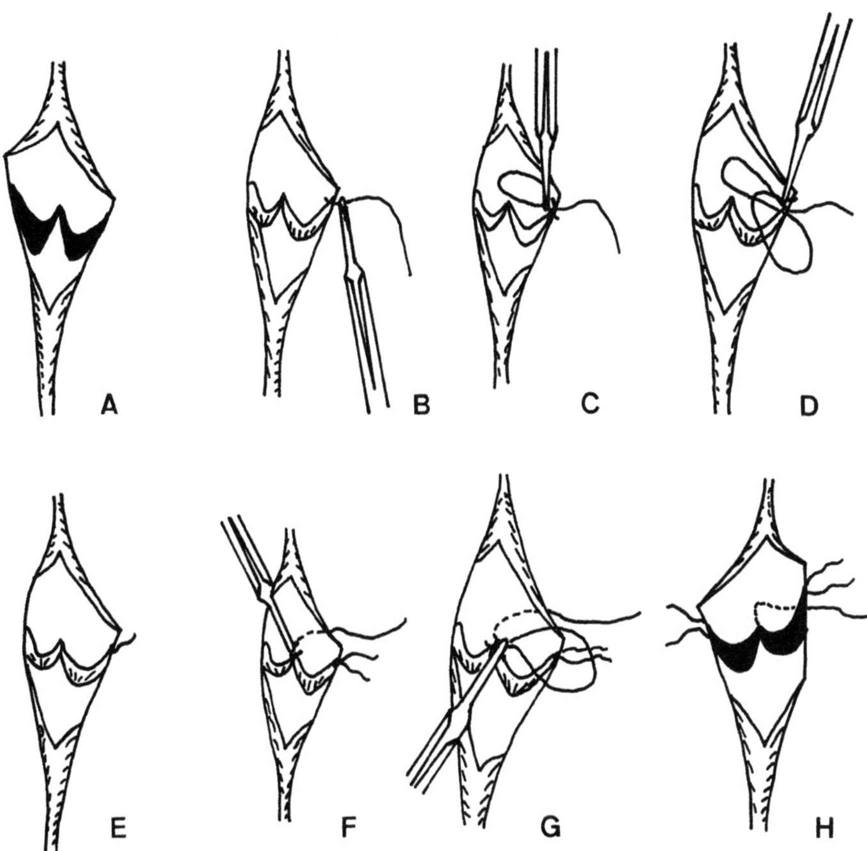

Fig. 89. Valvuloplasty according to Kistner, schematic representation. [Reproduced from drawings in Arch. Surg. *110*, 1338 and 1339 (1975).] For explanations see text

commissure. H: by applying similar sutures to the second horn of the valve, the valve is shortened by means of reefing sutures tied on the outside of the vein. The vessel may be closed.

As demonstrated in the drawing, the femoral vein is exposed over large distances for *Kistner's* operation. The entering veins are constricted by loops of thread, and the femoral vein is opened between holding stitches. Use of magnifying glasses or a surgical microscope is mandatory.

According to *Kistner* (1983), the procedure should be reserved for "primary valvular incompetence". For post-thrombotic states with widespread valvular incompetence he recommends grafts of valved venous segments. He reports 83% of good to excellent results for an average period of follow-up of 5 years. He stresses the necessity of ligating the perforators and interrupting the saphenous system. According to another report published in 1982, excellent results were obtained in 16 cases even without those additional measures (*Ferris* and *Kistner*).

Long-term control examinations by descending venography 2 to 13 years after operation revealed 15 of 17 such valvular reconstructions to have remained competent and merely 2 valves to have functionally deteriorated (3 and 6 years after operation).

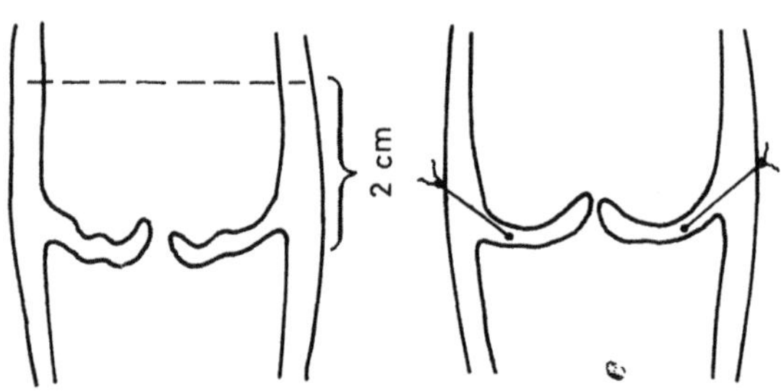

Fig. 90. Modified method of valve gathering. Transverse incision some 2 cm above the valvular agger visible from outside. [According to Raju, S.: Ann. Surg., vol. *197,* 688 (1983)]

The method of valvuloplasty according to *Kistner* was used by *Raju* (1983) in 20 cases. In one case local thrombosis occurred postoperatively. 17 other cases, in which phlebographic control was obtained, remained patent. In six cases, however, sudden occurrence of leg edema was observed during the postoperative period. Some of these incidents occurred as late as several years after the operation and in most of these cases the edema was transitory and its underlying cause unclarified. 16 extremities of 15 patients were reexamined. The complaints (ulceration, pain, edemata and recurrent phlebitis) largely subsided. All except two patients had regained full occupational capacity, and only one patient was wearing elastic stockings. Laboratory studies of circulatory parameters showed clearly improved reports in 10 patients, and in 15 extremities the repaired valve was proved to be competent by *Doppler* examination. 5 patients were subjected to phlebography postoperatively. In four of them no reflux or very mild reflux through the repaired vein was found, in one patient the reflux was unchanged as compared to the condition prior to operation. His hemodynamic parameters were found improved though, and leg ulcer had healed.

Raju modified *Kistner's* operational method in that he incised the vein transversely 2 cm above the valvular agger discernible from outside. Before and after valvuloplasty, the function of the valves was tested by flushing them with heparinous saline (Fig. 90).

Other valvuloplasties were performed by *Wall* (1979), *Schanzer* and *Price* (1982), *Fogarty* (1982) and *Huse et al.* (1983). *Wall* (1979) recommends the use of a surgical microscope.

Kistner, as well as others who used his method, are of the opinion that reconstruction of the valve alone will not suffice and that varicose veins, if present, and dilated and incompetent perforating veins must be removed by surgery.

e) Venous Ligatures in Cases of Post-thrombotic Syndrome

The post-thrombotic syndrome is caused by recannulation of the obliterated veins, after which a largely valveless tree of vessels is left behind. In peripheral leg veins the lack of appropriate luminal capacity for carrying off venous blood hardly matters – the dilatability of collateral veins seems to be sufficient. Recannulated leg vein thromboses instead cause damage by resulting in functional impairment of the peroneal muscle pump so that the venous pressure does not decrease or decreases only slightly during exercise. Previously, the pathogenesis of the post-thrombotic syndrome was explained by the increased hydrostatic pressure during standing. As mentioned on p. 153, this argument should not be rejected entirely, even though it was highly overrated in former years. Such overrating became evident in G. *Bauer's* recommendation to ligate the popliteal vein in order to cut off the column of blood and in this way reduce the venous pressure in the periphery of the leg. We would explain the lack of success of these ligatures by maintaining that collateral veins will dilate very soon after the operation and, thus, become incompetent as well.

The suggestion brought forward by *Linton* and *Hardy* (1948) seems to be more promising. They propose ligating the superficial femoral vein slightly below its confluence with the profound femoral vein, as the profunda usually has a competent valve slightly before the confluence. Besides, this vein is strong enough to be subjected to conditions of increased bloodflow without dilating to an extent that would result in valvular incompetence. This form of venous ligature has recently been found to be, at least temporarily, successful by *Kistner* and *Sparkuhl* (1979), and by *Queral* (1980).

Still, transpositions of valve-bearing veins or free transplantation of valved venous segments would seem to us to be more promising than venous ligatures.

C. Transplantation and Transposition of Veins

a) Historical Notes

The inventor of shunting arterial obliterations by autologous veins (venous bypass), *J. Kunlin,* was also the foremost pioneer in the field of veno-venous grafting. *Kunlin* observed progressive shrinking of the ring of anastomosis, which often led to long-term obliterations in veno-venous transplantions. For avoiding such obliterations, *Kunlin* recommended two methods: firstly, creation of a distal arterio-venous fistula (1954) and secondly, suspension of the anastomosing stitches to a wire ring surrounding the external circumference of the anastomosis (1974). Further progress was achieved by *Jacobson* and *Katsumura* (1965), who transplanted canine veins under the microscope, putting great emphasis on "extreme meticulousness of technique".

b) Endothelium-preserving Operational Method

Gottlob et al. (1975) proved that, as a rule, the greatest part of the endothelium is damaged during dissection of a long saphenous vein for the purpose of functional substitution of an artery by means of bypass. Regeneration takes 1 to 2 weeks. Such destruction of the endothelium may be largely avoided if during preparation the vein is touched neither by hand nor by swabs or instruments and unphysiologic tension is not exerted. In order to dissect the vessel, only the adjacent connective tissue may be seized by forceps and severed sharply from the vessel.

For purposes of transplantation endothelium-preserving dissection is not enough. The suture of the anastomosis must also be carried out in an endothelium-preserving manner. Investigations by means of en-face preparations showed that by conventional stitching of the anastomosis alone an approximately 10 mm wide cuff of the vessel is denuded of its endothelium on both sides of the suture. When two anastomoses are sutured, this will result in a denuded area of 4 cm length! Endothelium-preserving stitching of the anastomosis was achieved in the following manner: First of all, the severed ends of the vessel were prevented from shrinking, snapping together or conglutinating. It was found that searching for the conglutinated lumina alone caused endothelial damage of severest nature. In order to prevent snapping together and conglutination, the vessels were incompletely severed prior to full division. A holding suture was stitched through the edge of the opened vessel. By means of this holding suture the vessel was kept under constant tension. Using a specially devised frame, the transplant was extended to its approximate original length by means of the holding sutures. By clamping of the holding sutures to surgical towels, the ends of the host vessel were kept extended and conglutination of the lumina was prevented. During stitching, the vessels were adapted only by means of the holding sutures, and stitching was made without seizing the ends of the vessels by forceps. The procedure was first used for replacement of arteries by autologous veins. It was found that the endothelial cells obtained during dissection were still present, also maintaining their venous form in the artery for many months (*Gottlob et al.,* 1976). Degeneration of valves, as observed by *Weissenhofer et al.* (1974) 21 to 28 days after operation, did not occur under the conditions of the endothelium-preserving operational method.

We tried to use the endothelium-preserving method for venous replacement as well and found that progressive shrinking of the anastomosis, as observed by *Kunlin,* did not occur. Although there was some mild angiographic evidence of constriction in the region of the anastomosis during the first few days after the operation, these constrictions completely disappeared after several days or weeks. We assume that the region of anastomosis is always the site of depositing of thrombi during the early postoperative stage. If, however, the denuded area of the vessel extends over a short distance only (in histology 2 mm instead of 10 mm with conventional stitching methods), the parietal thrombus disappears soon and the vessel preserves its patency. This is the case even if systematic anticoagulation measures, creation of an arteriovenous fistula or suspension of the anastomosis are refrained from (*Gottlob,* 1977a, 1978a, 1982; *Gottlob* and *May,* 1977). Using the endothelium-preserving technique of dissection, we even succeeded in transplanting small neck veins of rabbits with good results (*Gottlob et al.,* 1982a).

Though it was not the purpose of these studies to ascertain the extent to which venous valves may be preserved by this method, we often observed that venous valves incidentally included in the transplant had retained their normal characteristics. Portions of free valvular cusps were found even in cases in which one horn of the cusp had been seized unintentionally by the suture.

We believe that we may recommend the endothelium-preserving operational method for all surgical interventions into the venous system. Its particular advantage is to be seen in the fact that creation of an arteriovenous fistula becomes superfluous.

Subsequent to our findings, similar conclusions were made by *Krupski et al.* (1977), and by *Bush* (1981). By applying the "gentle no-touch technique", *Bush* transplanted 20 venous segments in dogs, achieving patency in 90% of cases.

In view of repeated observations of valvular incompetence in experiments, *Raju* and *Perry* (1983) examined the valvular endothelium at different points of time after free transplantation of valved segments under the electron-microscope. They found progressive endothelial damage with a peak on the 28th day after the operation. By that time, the endothelium had disappeared almost completely. Basal membrane and collagenous framework were exposed. Four months after the operation, there was normal endothelium again. – Light-optical investigations of our own never revealed equally severe endothelial damage in the valvular area after transplantation. The question arises whether the authors used an endothelium-preserving technique in their operations. The text of their paper does not supply any clue to that.

Other authors (*Kistner*, 1975; *Huse et al.*, 1983) report that during valvuloplasty they "milk" the femoral veins in retrograde direction so as to evaluate the functionability of the valves in this region. – In view of our own experience we would be highly surprised to learn that such procedure does not cause massive endothelial damage.

Clamps also may be a decisive factor for the destiny of a valvuloplasty. *Gottlob* and *Zinner* (1962) observed complete desquamation of the endothelium in the region of compression by hard metal clamps. Milder damages were observed when the branches of the clamp had been covered by a rubber tube. The mildest endothelial damages were observed after use of a Hydagrip clamp according to *Fogarty* (*Gottlob and Rauhs*, 1978). One branch of this clamp is covered by a rubber tube filled with water.

A particularly gentle method of temporarily occluding the lumen of a vein is described by *Gottlob et al.* (1982): The vein is circumstitched by a large curved needle in such a manner that ample quantities of connective tissue or muscle tissue are left between the vessel and the suture (approximately 2–0). The two ends of the circumstitched suture are not tied up but pulled through a thin plastic tube of some 2 cm length and fixed by clamp to the distant end of the tube after constriction of the vessel carried out by pulling of the threads. *Bylsma* (1985) alternately closed veins to be used for a bypass by means of metal clips and sutures. The frequency of occlusions was increased in the constricted vessels.

By evaluation of the prostacyclin production, *Bush et al.* (1984) investigated into the question of what the best method of preserving the vitality of venous segments destined for transplantation was. They pleaded for storage in body-warm solutions. Observations of our own, according to which the vitality of veins may be preserved by quick deep-freezing, are in opposition of their findings (*Gottlob et al.,* 1982b).

c) Autologous Transplantation of Valve-bearing Venous Segments

1. Experiments

Transplantations of venous segments are based on the pioneer work of *Kunlin* (1954) and *Jacobson* and *Katsamura* (1965) as well as on experimental studies conducted by *Scott* and *Dale* (1963), *Haimovci et al.* (1963, 1970), and *Carino et al.* (1964).

Investigations into the possibilities of valve transplantations first of all showed that valves are transplantable and may retain their functionability unless early-stage thrombosis occurs. Early-stage thrombosis tends to be overlooked in experiments, due to the fact that in laboratory animals spontaneous thrombolysis occurs almost regularly and, after intervals of several days or weeks, misleads the observer to the erroneous assumption that the transplant has been patent without interruption (*De Weese* and *Niguidula,* 1960). Early-stages thrombosis is caused by the low velocity of current in the venous bloodflow (*Kunlin,* 1954; *Dost,* 1965; and others) as well as by the fact that the intima is largely denudated of thrombi-repellent endothelial cells (*Gottlob* and *May,* 1977). For avoiding early-stage thrombosis in valvular grafts, treatment by anticoagulants during the operation and for several days thereafter (*McLachlin et al.,* 1965; *Taheri et al.,* 1982; *Waddell et al.,* 1967) and construction of a distal arteriovenous fistula for accelerating the bloodflow (*Aschberg* and *Hindmarsh,* 1971) are recommended. After transplanting valved venous segments in dogs, *Kroener* and *Bernstein* (1981) made the following observations:

In a group of ten dogs that had not received concomitant treatment, there was angiographic evidence of early-stage thrombosis, marked stenosis or filling defects in 90% of the cases. Some time later, however, 75% of the grafted valves were competent. – In a second group with distal arteriovenous fistula all transplants remained patent, 86% of them had competent valves. Finally, in a third group consisting of 10 animals, all transplants remained patent. In that group both arteriovenous fistulae had been constructed and postoperative treatment by dextran infusions had been given.

Investigations of our own (*Gottlob* and *Kimmel,* 1973) seem to indicate that once thrombosing has occurred in a venous segment, the valves contained in it will show a tendency toward degeneration after spontaneous lysis. Shrinking and thickening of the valvular rudiments may, however, take a considerable amount of time. *Kroener* and *Bernstein* advocate a similar view with respect to transplanted valves, and the experimental studies conducted by *Tompa et al.* (1975) also support the assumption. Differing views are held by *Aschberg* and *Hindmarsh,* who, in a small number of cases, found valves that were intact and apparently competent some time after early-stage thrombosis. *Baird et al.* (1964) support the opinion according to which early-stage thrombosis need not necessarily result in destruction of the valves. In contrast, *De Weese* and *Niguidula* (1960) invariably found incompetent valves after recannulation of thrombi.

2. Clinical Experience

Haimivici (1970) gives a survey of less recent attempts at replacement of obliterated portions of veins.

Transplantations for the purpose of valve reconstruction were performed by *Taheri et al.* (1982) and *Raju* (1983).

Taheri et al. dissected 2 cm long valved segments of the axillary vein and implanted them into the femoral vein. A total of 23 extremities of 20 patients was treated. Procedures that might have feigned operational success were not carried out later than 1 year before the transplantation. As from the day of operation, heparin was administered in the form of continuous infusion, and anticoagulant treatment was continued to be administered orally for 3 more months. In addition, active and passive leg exercises were carried out postoperatively.

Despite markedly improved clinical findings, 9 patients failed to show improved venous pressure readings either at rest or during exercise. According to the authors, it may be possible that reconstruction of the uppermost valve alone is not sufficient and because of that in future several venous valves will have to be transplanted at a time.

By a similar method *Raju* transplanted 22 venous valves. The immediate results were quite promising: 9 crural ulcers were healed, in two cases the edemata were improved, and in two patients the pain largely subsided after the operation. In 10 out of these 14 extremities improved laboratory findings concerning venous hemodynamics were obtained. In three cases no such improvement was observed despite subjective amelioration, and 8 cases were dismissed as surgical failures. In one case the failure was attributed to a technical error, i.e. an axillary vein of insufficient size was grafted, and in eight cases to reflux in the grafted vein.

After free transplantation of axillary vein segments, *Rushton* (1982) observed subsequent incompetence of the transplants in "alarming numbers". The failure was attributed to dilation of the transplant. As a preventive measure, *Rushton* has resorted to transplanting two segments stitched together in the form of a double-barreled gun. It is assumed that the two lumina will better cope with the heavy bloodflow from the leg vein. In several transplantations *Raju* sheathed the grafted segment in a Dacron cuff in order to prevent dilation. Apparently he used the porous material available for arterial substitution.

d) Homologous Veno-venous Transplantation

So far homologous grafts to the venous system have only been performed experimentally. *Kojima et al.* (1982) investigated to which extent rejection may be prevented by immunosuppression. The authors proceeded as follows:

6 dogs received an autologous transplant of 3 cm length into the femoral vein. 7 dogs received a homologous transplant without immunosuppression, and 7 dogs received a homologous transplant with immunosuppression by Azathiaprine and Prednisolon.

After 4 weeks the autotransplants were patent. The homologous transplants without immunosuppression were invariably thrombosed, whereas no more than 25% of the animals that had received immunosuppression showed obliterations.

Gaudini et al. (1983) investigated to which degree rejection may be prevented by administration of aggregation inhibitors. Their investigations also involved canine femoral veins. With and without administration of the platelet-effective substance, all cases invariably showed obliteration of the transplant after 18 ± 7 days.

Experimental homologous transplantations of venous valves were carried out by *Waddell et al.* (1964) and *McLachlin et al.* (1965). In no case was prolonged functionability of the transplanted valve observed.

All experimental results mentioned above seem to indicate that, for the time being, homologous transplantations are not suitable for clinical use. Moreover, suitable axillary vein segments are available for valvuloplasty as a rule. The removal of short segments of vessels from the axillary vein does not cause any complaints in the upper extremity.

For purposes of valvuloplasty in patients with venous incompetence, *Ackroydt et al.* investigated in which manner the mechanical properties of the vessel change when the segment is preserved by glutaraldehyd and sterilized by formalin afterwards. They found that neither breaking strength nor expandability had changed to any significant extent as a result of such treatment. – It must be added in this respect, however, that experience seems to indicate that in peripheral veins replacement of a segment with or without valves is possible only if the endothelium is vital and largely intact. Merely under conditions of arterial flow may intactness of endothelium be dispensed with!

e) Transposition of Veins

1. Femoral Bypass According to Husni (1970) and May (1972)

As the drawing shows, almost simultaneously with *Husni*, Cleveland (1970), *May* had the idea of replacing femoral veins with severe post-thrombotic changes by implanting the long saphenous vein into the lowest segment of the popliteal vein, In this procedure,

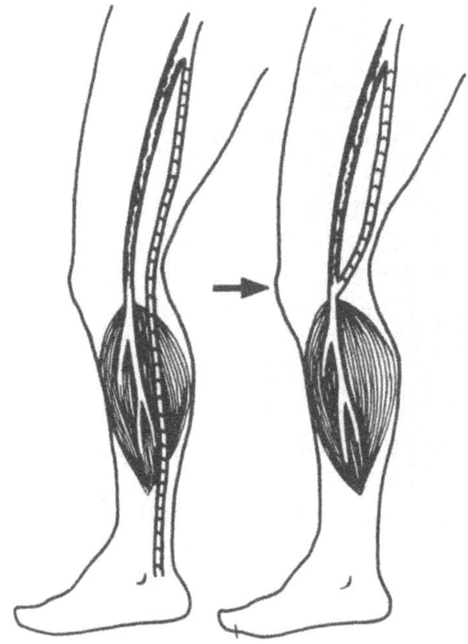

Fig. 91. Schematic representation of femoral bypass

the long saphenous vein, the valves of which are completely competent, is left continually attached to its surrounding tissues (Fig. 92).

Later – sometimes as soon as 2 years afterwards – the long saphenous vein degenerates in a slightly varicose manner and its valves are destroyed in this process (Fig. 93).

The fact is noteworthy. We shall have to pay close attention to long-term control examinations when attempting to graft valve-bearing segments of veins in man.

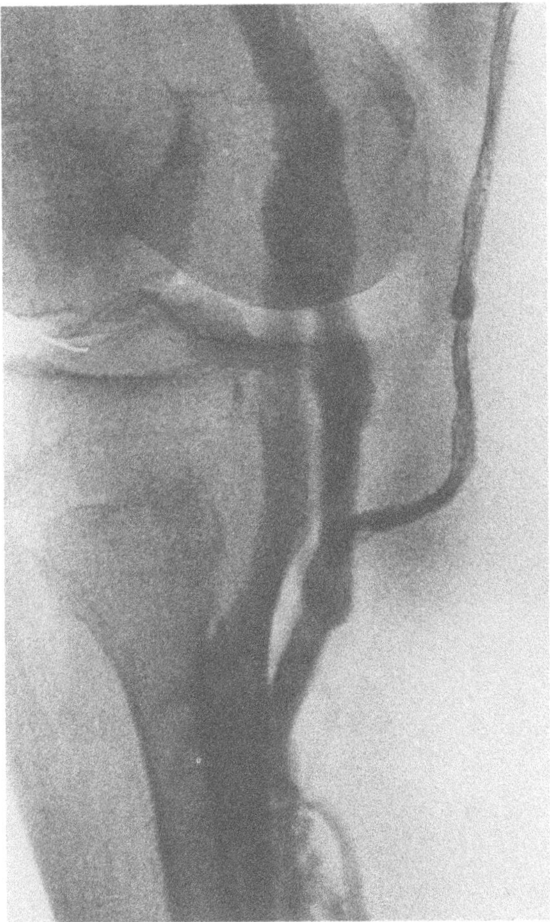

Fig. 92. Femoral bypass. Control film obtained one year later. The valves of the internal saphenous vein are still competent

As early as in 1954, *Warren* and *Thayer* first attempted to transpose the long saphenous vein into deeper layers of the thigh under conditions of the post-thrombotic syndrome. *Katzenstein* reported on similar surgery as early as in 1911. In some of their cases *Warren* and *Thayer* performed an anastomosis to the popliteal vein in the distal region so that they may be said to have applied a procedure fundamentally similar to that designed by *May* and *Husni*.

2. Transposition of Femoral Vein in Proximal Thigh

This method also goes back to *Kistner* (*Kistner* and *Sparkuhl*, 1979). If the valves in the femoral vein below and above the entrance of the profunda were incompetent, the femoral vein was divided below the

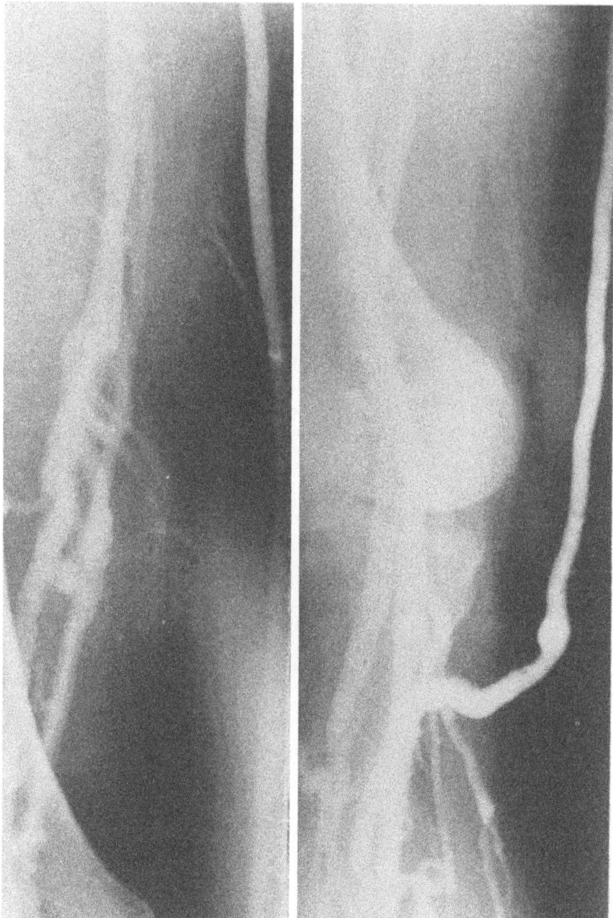

Fig. 93. Control film of femoral bypass, obtained 5 years later. The internal saphenous vein has meanwhile become incompetent. (This case is not identical with the case in Fig. 92)

entrance of the profunda and anastomosed end-to-side either to the long saphenous vein or to the profound femoral vein. The operation aims at interposing a valve into the outflow from the superficial femoral vein and in this way preventing reflux (Figs. 94C and D). In 1982, *Ferris* and *Kistner* reported on 14 cases of venous transpositions of the described nature. Reexaminations were made for periods of up to 6 years. 11 of the 14 patients showed excellent or good results. One patient was unimproved, and two patients developed recurrent ulcers 3 years after the operation. In one of these cases the transposed valve was proved to be competent by venography. According to the authors, optimal results may be achieved by a combination of vein transposition and ligatures of incompetent perforating veins. Vein transposition alone often is somewhat less successful than the combined method. Ligature of perforating veins alone brings short-lived success only. A further disadvantage of it is a considerably high rate of recurrence.

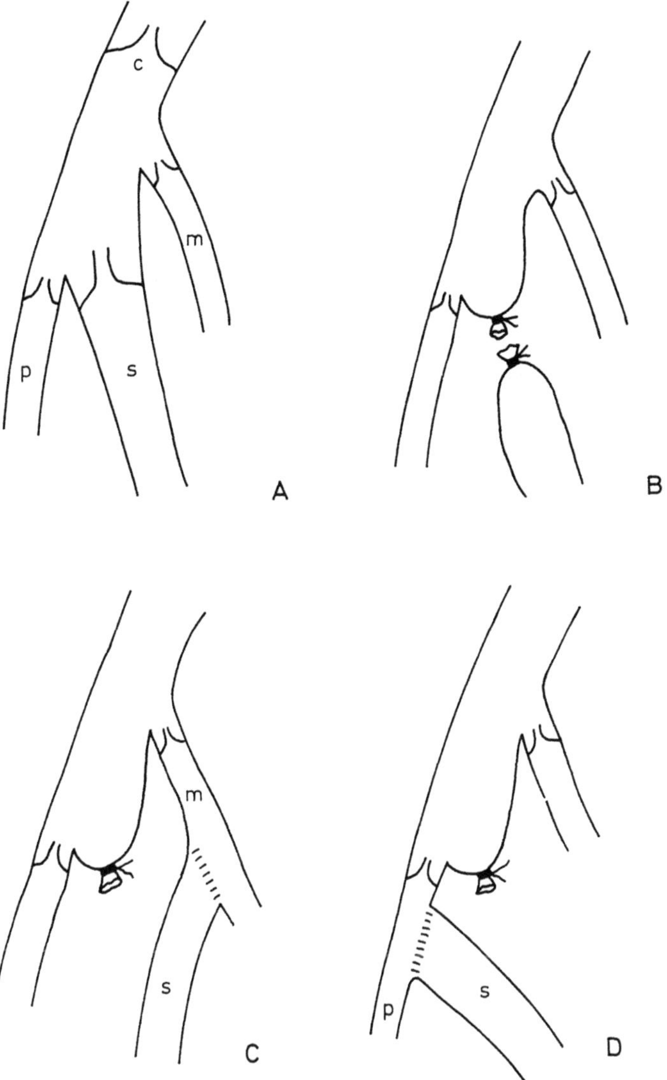

Fig. 94. Venous ligature and venous transposition in proximal thigh, schematic representation. **A** Normal situation. **B** Ligature of femoral vein. **C** Terminolateral implantation of (superficial) femoral vein into long saphenous vein. **D** Terminolateral implantation of (superficial) femoral vein into deep femoral vein. *C* common femoral vein. *S* superficial femoral vein. *P* deep femoral vein. *M* long saphenous vein

Queral et al. (1980) performed venous transpositions in the thigh, if the femoral vein showed valvular incompetence. In 10 of their cases they found an internal saphenous vein in which the uppermost valve was intact. Therfore, the femoral vein was severed and its peripheral end was anastomosed to the long saphenous vein distal to the uppermost valve. In two other cases an incompetent femoral vein was implanted into the deep femoral vein, distal to a competent valve slightly before the junction.

All cases reported by *Queral et al.* were characterized by preoperative evidence of bidirectional bloodflow in the femoral and popliteal veins, established by *Doppler* readings. Postoperatively, nine extremities out of 12 showed unidirectional bloodflow in the femoral vein, and five unidirectional bloodflow in the popliteal vein. – Before the operation, pressure recovery time after exercise was pathologically reduced in all extremities operated on (average level 10.7 sec.), whereas after the operation no more than 3 extremities showed pressure recovery time readings below 20 sec. (threshold level of pathology). The average pressure recovery time had increased to 21.9 sec. The patterns of photoplethysmographically measured recovery times were almost identical to those obtained by venous pressure measurement. *Queral et al.* healed the ulcers, their patients were suffering from by conservative methods *prior* to venous surgery.

In 1981, *Johnson et al.* reported on 14 extremities which were treated by venous transposition in the proximal thigh. In 9 out of 12 legs, pressure recovery time readings improved temporarily but reached preoperative levels after 12 to 18 months. In no more than 3 cases was improvement of venous incompetence achieved. Four venous ulcers recurred postoperatively. These results were considerably less successful than those obtained by *Kistner*. *Johnson et al.* attributed this to the fact that in a majority of his patients *Kistner* had ligated the perforating veins simultaneously with the transposition. Another reason *Johnson et al.* gave was that in their own patient material the valves had been in a state of destruction down to the periphery. Furthermore, they pointed out the inaccuracy of evaluating valvular incompetence by phlebographic means.

Other authors reporting on transposition of veins for treatment of venous incompetence are *Schanzer* (1982), *Taheri* (1982), *Huse* (1983) and *Raju* (1983). *Raju* operated on 3 patients, all of whom showed improvement immediately after surgery. In one of these patients the improvement lasted for only 2 years, after which there was recurrence of congestive dermatitis as well as positive evidence of reflux in venography. This patient was then subjected to free axillary vein transplantation, the results of which were good on a short-term basis.

For prevention of dilation, the transplant was sheathed in a Dacron cuff. *Sandmann et al.* (1982) reported on four cases of femoral vein transposition. Three patients were free from complaints, 1 patient was clinically improved.

3. Operation According to Palma

The methods described above aim at the improvement of complaints caused by valvular damages in the region of the leg veins. In the veins below the inguinal ligament, obliterations cause complaints during the acute phase only. After the end of the thrombotic process, the patient passes through an interval during which he is relatively free from complaints. His complaints will reoccur massively when the obliterated vessels become devoid of valves due to recannulation. The patient will suffer from "reflux disease" then. Quite a different situation presents itself in thrombosis of the pelvic veins. In a majority of cases these veins are valveless even in phlebologically healthy persons. Recannulation is observed only rarely. The clinical picture is determined by the extent of obliteration. The collateral veins are not sufficient to replace the wide lumen of the pelvic veins, due to the fact that flow resistance is determined by the radius to the power of four. Surgical therapy of the status after pelvic vein thrombosis must aim at creating new vascular capacity for carrying off venous blood. – In view of the fact that this is not a valvular problem, the operation according to *Palma* will be only briefly described for the sake of completeness.

A schematic representation of the operation is given in Fig. 95. For the purpose of shunting an obliterated iliac vein, the counterlateral internal saphenous vein is rerouted suprapubically through a skin tunnel and anastomosed to the femoral vein on the side affected by the disease.

May gave a detailed description of this operational technique in 1972. The results may be considerably improved by applying an endothelium-preserving technique (*Gottlob* and *May,* 1977). The decisive factor for the success of the operation is to be seen in differences of pressure between the healthy and the obliterated venous systems both at rest and, in particular, during exercise. The transposed saphenous vein subsequently dilates, however (Fig. 96), the valvular status in the transposed vessel is irrelevant.

Provided that the pressure-difference between the affected and the unaffected side is large enough, arterio-venous anastomoses, as recommended by some authors, such as *Vollmar* and *Hutschenreiter* (1981), may be dispensed with when operating in an endothelium-preserving manner.

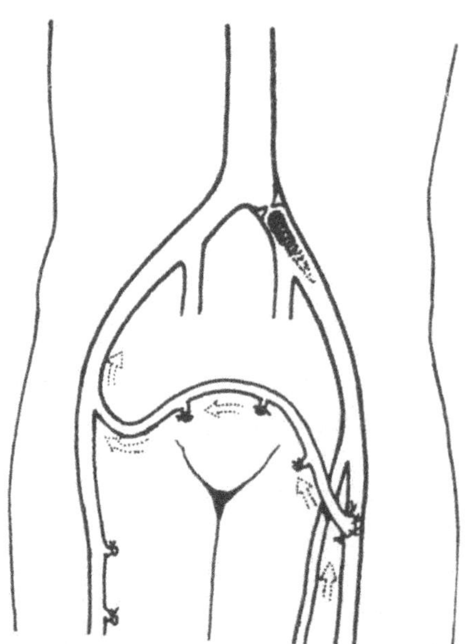

Fig. 95. Schematic representation of Palma's operation

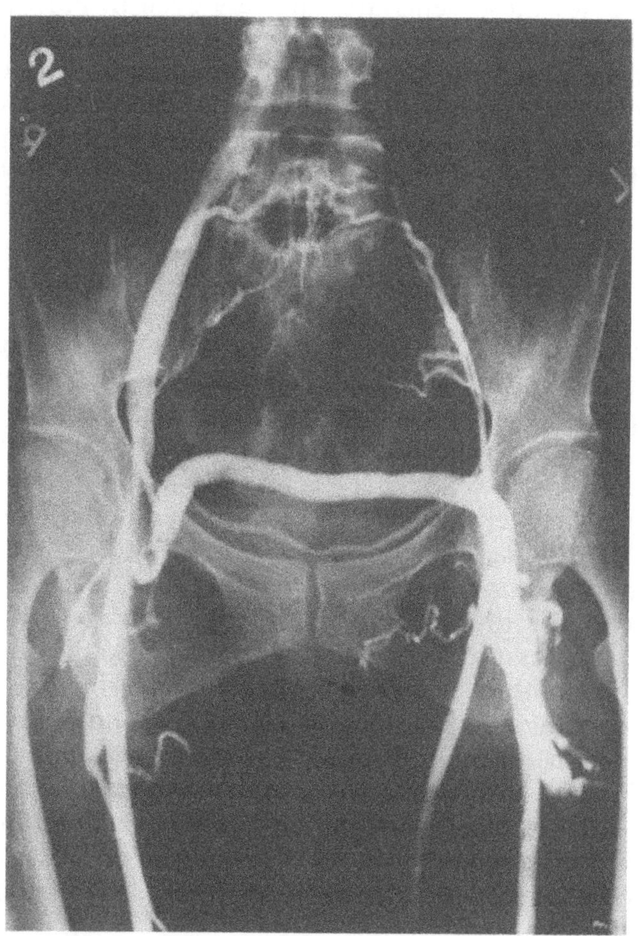

Fig. 96. Control film showing status post Palma's operation five years after surgery

D. Critical Discussion of Methods of Replacing Venous Valves

a) General

According to *Linton* (1953), chronic ulceration of the lower extremities caused by venous thrombosis is a condition that has afflicted humankind ever since man assumed an erect posture. Attempts aiming at improving that condition by means of surgery are well justified and desirable. Such surgical attempts, however, may promise success only if an extremely critical attitude is chosen. So, first of all, we would raise the following question:

What are the fundamental expectations we may have in valvuloplasty?

Studying the functions of venous valves, we find that their most important function is that of supporting the muscle pump. During the "systole" of the muscle pump, i.e. when the veins are contracted, blood is drained from the deep veins and passes on in a centralward direction. Venous valves are not necessarily required for that purpose. The direction of the bloodflow is ensured by the fact that there is less resistance to the current in the central part of the system than in the periphery. In the absence of valves, however, the "diastole" of the muscle pump results in reflux of some part of the blood previously ejected. Such reflux causes the muscle pump to work inefficiently – resulting in edemata and other manifestations of post-thrombotic syndrome. – By valvuloplasty in a far more proximal region we may hardly be able to counteract that mechanism. This explains the results obtained by *Johnson et al.* (1981), who failed to observe improved venous pressure readings despite the presence of functional valves.

In addition to the *hydrodynamic* function of venous valves as explained above, they have a *hydrostatic* function. When the subject is in a standing position and the peripheral venous pressure is low, as it is observed to be shortly after exercise, the phase of exercise is followed by reflux to the periphery. "Pressure recovery time", i.e. the period of time that passes until the pressure reattains the level prior to exercise, is markedly shortened. – This hydrostatic function doubtlessly may be improved by valvuloplasty in the inguinal region. One must keep in mind, however, that this is no more than a partial function and that a single valve alone is not able to substitute the entire valvular apparatus of the leg, consisting of numerous valves. Besides, we are in no position yet to judge with certainty the relative value of the hydrostatic function as compared to that of the hydrodynamic function. In order to judge that relative value for a given extremity, we would also have to take into consideration to which extent valves have been preserved in the periphery, e.g. in muscle veins.

In this connection the observations of *Shull et al.* (1979) are noteworthy. The authors observed that, under conditions of occlusion of the femoral veins with or without involvement of the iliac veins, crural ulcers do not occur unless the valves of the *popliteal vein* are incompetent.

b) Assessment of Clinical Success

When assessing the extent to which subjective or even objective improvement after valvuloplasty may be attributed to the operation itself, we should apply particularly critical standards. If, simultaneously with valvuloplasty, other measures were taken, it is impossible to assess the successfulness of the operation. Under conditions of servere varicosis, the deep veins must carry off physiologic quantities of blood and, in addition to that, blood that had flown back through the private circulation (see p. 129). If, simultaneously with valvuloplasty, high saphenous ligature is performed along with stripping, the private circulation is eliminated so that the quantity of blood to be carried off by the deep veins has been considerably reduced. This also results in a reduction of dilation so that the valves regain their capability of closing.

Ligature of incompetent perforating veins is of particular importance in this respect. Other than is the case with primary varicosis, the function of these vessels has been definitely clarified as far as the post-thrombotic syndrome is concerned. The bloodflow is directed from the deep system to the surface. By ligature of perforators, *Schanzer et al.* (1982) succeeded in achieving substantial improvement in 23 patients. During observation periods of up to 50 months, no more than 2 cases were found to be recurrent.

Other *measures carried out concomitantly with the operation* may also lead to misinterpretation of operational success. Crural ulcers often heal promptly after a few days of bedrest. An elastic bandage or a changed way of life – e.g. avoiding prolonged standing – may have miraculous effects. We must also point out to the *placebo effect,* which often works amazingly in patients suffering from the post-thrombotic syndrome. We have observed in quite a number of cases that complaints caused by this syndrome were markedly improved by totally unspecific operations of the leg!

c) **Critical Discussion of Valvuloplastic Methods**

1. Obsolete Methods

The substitute valve according to *Psathakis,* the sheathing method according to *Hallberg* and the invagination method according to *Eiseman* and *Malette* did not find general acceptance.

One argument against the substitute valve is that by interposing the gracilis tendon between artery and vein in the knee region, the vein is constricted during the systolic phase of the muscle pump, i.e. during the very phase in which normally valves are open under physiologic conditions.

Invaginations, as a rule, are not well tolerated by vessels and result in local necrosis or sclerosis.

Invagination of the venous wall in order to form a new valve leads to cusps whose thickness is twice that of the venous wall. It is true though that *Malette* (1981) reported on findings according to which the invaginated portion of the wall loses media and adventitia so that a very delicate cusp covered by endothelium is left. *Bergan* (1981), however, did not succeed in reproducing the results of *Eiseman* and *Malette.*

2. Valvuloplasty According to Kistner

Assessment of this method is particularly difficult. First of all, those general objections that were also raised in connection with other methods (p. 201 f.) must be referred to once more. In addition, we would pose the following questions:

1. How often do we find *isolated slackening of the cusps* of the uppermost valve of the femoral vein causing valvular incompetence? According to *Fogarty* (1982), 1 patient out of 15 is suited for valvuloplasty. According to *Taheri et al.* (1982), his patient material mostly showed valves in the respective regions that were thickened, changed by inflammatory processes or adherent. In contrast to *Taheri et al., Raju* (1983) reported on patient material consisting of a large number of cases in which incompetence of the deep veins *without* signs of previous thrombosis was found (see also pp. 155 f.). – This inevitably leads to the assumption that various authors are reporting on vastly differing patient material!

2. Will reconstruction of the *uppermost valve of the femoral vein* suffice to eliminate incompetence of the deep veins? This question is answered in the negative by several authors. In particular they point out the fact that in many cases the venous pressure during exercise and the pressure recovery time as well as the refilling time in

plethysmography were not improved after valvuloplasty (*Taheri,* 1982; *Schanzer,* 1982; *Eriksson et al.,* 1983). According to *Shull et al.* (1979), leg ulcers occur only if there is valvular incompetence in the popliteal vein. Besides, it is hardly imaginable that the uppermost valves of the femoral vein might contribute toward supporting the muscle pump. The frequently reported absence of improved venous pressure parameters may be explained by continued insufficiency of the muscle pump. Besides, observations made by several authors (*Luke,* 1943; *Hogensgard,* 1943; *De Camp,* 1951; *Fell,* 1975; *Johnson et al.,* 1981; Yao, 1982; *Raju,* 1983) seem to indicate that radiologic evidence of valvular incompetence with marked reflux may be found in persons that are completely free from any symptoms.

3. To which extent do delicate valvular structures tolerate *surgical manipulation?* According to our own experience, the reaction of valves to mechanical trauma is deposition of thrombi, which may result in subsequent shrinking. In all likelihood the valvular function remains intact under conditions of the endothelium-preserving operational method only (pp. 188 ff.). In view of the fact that histologic reports on the status of valves after suturing are not available, we conducted *experimental investigations* into that question:

Under anaesthesia, the jugular veins of rabbits were exposed at the confluence of the anterior and posterior facial vein. Both facial veins regularly show well-developed valves shortly before their confluence. After incision of the jugular vein, a valve was exposed and incised by a pair of fine scissors. The incision was then sutured by 10–0 atraumatic stitches. The incision of the jugular vein was also stitched. – 4 days after the operation, the vein was re-exposed. Whole-thickeness specimens of the valvular region were made (for method see pp. 43 f.). After inspection of the whole-thickness preparations, paraffin sections were made. In addition to 2 short-term observations (4 days after operation), long-term observations (3 months after operation) were made by means of the same method.

Results: In both short-term observations (4 days) the en-face whole-thickness preparation showed that the area of suturing had already been covered by endothelium (Figs. 97 A and B). The cusps were free. – In the paraffin sections moderate thrombotic deposits in the area of stitching and infiltrations of leucocytes in the vicinity of the stitches were found (Fig. 98).

In three long-term observations, 3 months after the operation, the cusps were found to be adherent to the parietal venous wall.

Our conclusions from these observations are: In the valvular region, sutures may be applied with relatively modest tissue reaction. The areas of suturing are covered by endothelium within a short time.

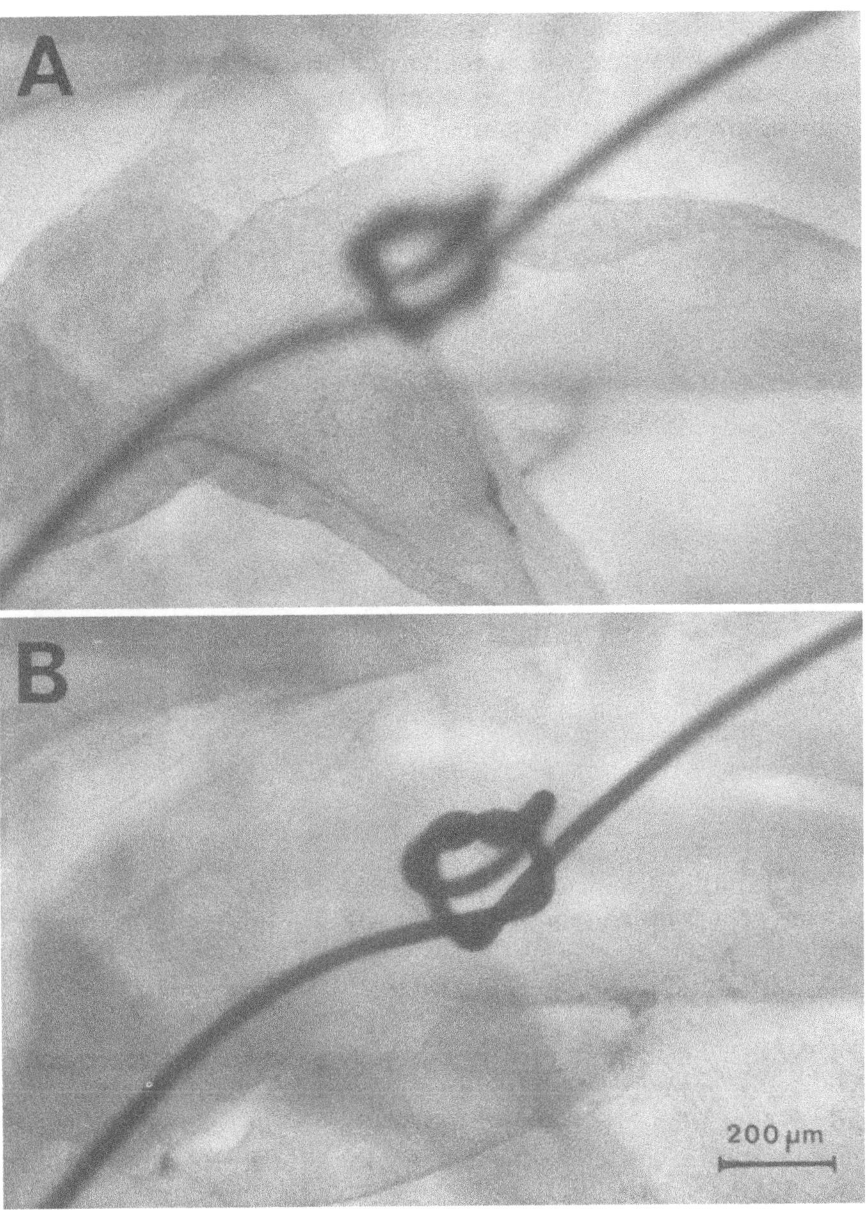

Fig. 97. Canine jugular vein. A valvular cusp was incised and stitched by 10-0 suture. **B** Deep level showing parietal endothelium. **A** The same area shown on valve level. The valve is noticeably covered by a faintly stained endothelium. Whole-thickness preparation, silver stained

There are modest deposits of thrombi. – Our studies, which had to be discontinued for unrelated reasons, do not allow drawing general conclusions from the long-term observations made. At least with valves of rabbit veins, which are particularly delicate, there seems to be a certain risk that on a long-term basis additional damages might occur in the region of the sutured cusp.

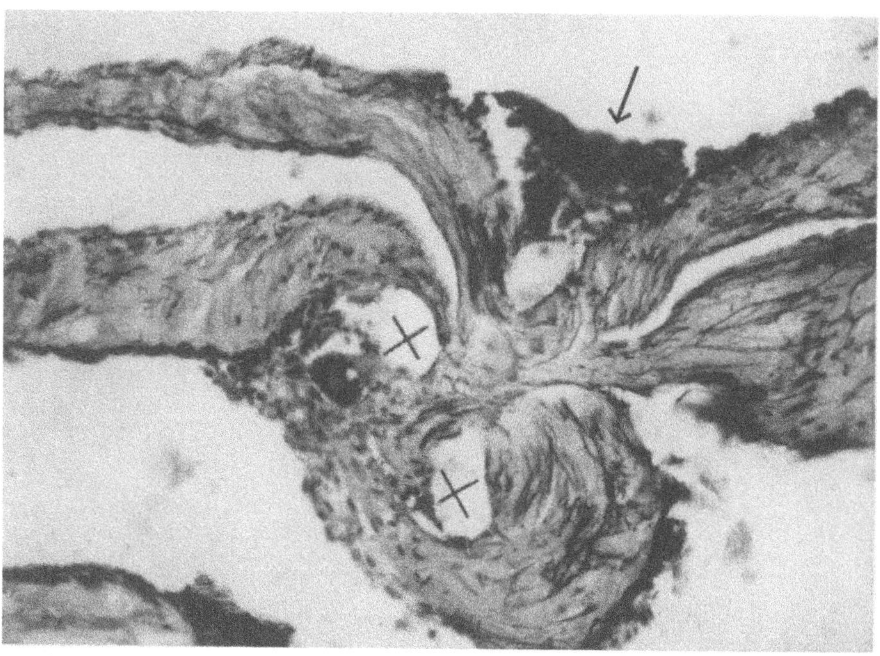

Fig. 98. Paraffin section. 4 days after suturing of the valve. The figure shows the same area as the en-face preparation (Fig. 97). Infiltration of leucocytes clearly visible in the area of the stiching *(x)*. There are also thrombotic deposits (arrow)

4. With what degree of certainty may valvular incompetence be assessed in *exposed veins?* The investigations of *Edwards* (1936) have shown that the cross-section of veins is usually elliptic. This configuration is caused by the fact that the vein is pressed towards the fascia by the skin or, in the case of deep veins, toward the bone by fascia and muscles. Venous valves are arranged in such a manner that the free edges of the cusps are parallel with the direction of the longest diameter of the ellipse (see p. 32).

Exposed veins tend to assume circular cross-sections. In that case the edges of the cusps are less tense. Therefore, the valve may be incompetent in this situation due to the fact that, by elimination of the elliptic outline, the diameter of the cross-section is shortened. This should be taken into consideration when attempting, as recommended by various authors, to assess valvular competence in exposed vessels.

d) Critical Assessment of Transplantation and Transposition

With respect to these methods, the same objections may be brought forward as have previously been raised in connection with valvuloplasty. Still, the advantage of these methods is to be seen in the fact that the valvular cusps need not be touched. According to *Raju* (1983), we must assume, however, that the success achieved by valvuloplasty will last longer than that of transplantation and transposition. This may be due to the following reasons:

1. Occurrence of valvular incompetence in venous segments used for valve substitution – possibly caused by a generalized tendency toward qualitatively impaired valves in the individual concerned and

2. dilation of the transplanted or transposed vein due to increased bloodflow.

3. We would refer to the findings of *Raju* and *Perry* (1983, see p. 190), according to which transplanted valves show endothelial damages, which reach their peak around the 28th day after the operation. Such long-term damages may, of course, lead to thrombosis or at least adhesion, thickening or shrinking of cusps and, thus, to valvular incompetence.

Summary – Part V

Venous valves in veins used for substitution of arteries by way of reversing of the vessel may contribute to improved flow conditions by absorbing reflected pressure waves. On the other hand, they may cause stenosis. Operations of the venous system should be carried out in an endothelium-preserving manner. Missing valves in the femoral vein may be replaced by free transplantation of valve-bearing axillary veins or transposition of the femoral vein. Slackened cusps were successfully corrected by valvuloplasty according to Kistner. Ultimate assessment of the methods used for substitution of valves is not yet possible. Further improvement of methods is desirable.

References

Abramowitz, H. B., Queral, L. A., Flinn, W. R., Nora jr., P. F., Peterson, L. K., Bergan, J. J., Yao, J. S. T.: The use of photoplethysmography in the assessment of venous insufficiency. Surgery *86*, 434 (1979).

Ackroyd, J. S., Pattison, M., Browse, N. L.: A study of the mechanical properties of fresh and preserved human femoral vein wall and valve cusps. Brit. J. Surg. *72*, 117 (1985).

Ahberg, N. E., Bartley, O., Chidekel, N.: Retrograde contrast filling of the left gonadal vein. Acta Radiol. (Diagnost.) *3*, 385 (1956).

Ahberg, N. E., Bartley, O., Chidekel, N., Fritjofson, Å.: Phlebography in varicocele scroti. Acta Radiol. (Diagnost.) *4*, 517 (1966).

Ahberg, N. E., Bartley, O., Chidekel, N.: Right and left gonadal veins. An anatomical and statistical study. Acta Radiol. (Diagnost.) *4*, 593 (1966 a).

Albrechsson, U., Olsson, C. G: Thrombosis after phlebography a comparison of two contrast media. Cardiovasc. Radiol. *2*, 9 (1979).

Alison, D., Rouleau, P., De Cassin, P., Dupuy, J. C., Betrand, P.: Study of low-iodine-concentration contrast material (Ioxaglate 20%) versus Ioxaglate 32% in phlebography of the lower limbs. Joint Meeting of the European Society of Cardiovascular and Interventional Radiology and the European College of Angiography, Wien, 22.–27. April 1985.

Aquapendente *see* Fabricius, H. ab Aquapendente.

Arenander, E.: Hemodynamic effect of varicose veins and result of radical surgery. Acta Chir. Scand. Suppl. 260 (1960).

Arnoldi, E. C., Haeger, K.: Ulcus cruris varicosum-crux medicorum. Lakartidningen *64*, 2149 (1967).

Aschberg, S.: Crural venous obstruction or incompetence. Acta Chir. Scand. Suppl. 436 (1973).

Aschberg, S., Hindmarsh, T.: Distal arterio-venous shunts in autotransplantation of canine venous valves. Acta Chir. Scand. *137*, 503 (1971).

Astrup, T., Glas-Greenwalt, P., Dalton, B. C., Caulfield, J. B., Mundt, E. D.: Localization of fibrinolytic activity at the lining of valves from heart and veins. Amer. Heart Assoc. Meeting, November 1971 (Abstract) (1971).

Baba, H., Djordjevic, M., Kiso, J., Hamada, O., Moskowitz, M., von Recum, A.: Hemodynamic effects of venous valves in aorto-coronary bypass grafts. J. Thorac. Cardiovas. Surg. *71*, 774 (1976).

Baird, R. J., Lipton, L. H., Miyiagishima, R. T., Labrosse, C. J.: Replacement of the deep veins of the leg. Ann. Surg. *89*, 797 (1946).

Bardeleben, K. v.: Das Klappendistanzgesetz. Jenaische Z. Naturwiss. *14*, 467 (1880).

Barnes, R. W., Garrett, W. V., Hummel, B. A., Slaymaker, E. E., Meixner, W., Reinertson, J. E.: Photophlethysmographic assessment of altered cutaneous circulation in the postphlebitic syndrome. Proceedings, A.A.M.I. 13th Annual Meeting, Washington, March 28 – April 1 (1978).

Barnes, R., Ross, E., Strandness, D. E.: Differentiation of primary from secondary varicose veins by Doppler ultrasound and strain gauge plethysmography. Surg. Gyn. Obstet. *141*, 207 (1975).

Basmajian, J. V.: The distribution of valves in the femoral, external iliac and common iliac veins and their relationship to varicose veins. Surg. Gyn. Obstet. *95*, 537 (1952).

Bassi, G.: Perforantes – Diszision mit der Hakenmethode, Technik, Indikationen, Ergebnisse. In: Venae perforantes (May, R., Partsch, H., Staubesand, J., eds.). München-Wien-Baltimore: Urban & Schwarzenberg. 1981.

Beesley, W. H., Fegan, W. G.: Investigation into the localization of incompetent perforating veins. Brit. J. Surg. *57*, 30 (1970).

Bergan, J. J.: Venous reconstructive surgery. Surg. Clin. *62*, 239 (1982).

Bergan, J. J.: In discussion to A. F. Johnson et al. Arch. Surg. *116*, 1466 (1981).

Berutsen, A.: An varicer paa und extremiteterne. Kopenhagen: Brauner. 1925.

Berutsen, A.: Die Varizen der unteren Extremitäten mit besonderer Berücksichtigung ihrer Ätiologie und der chirurgischen Behandlung. Acta Chir. Scand. *62*, 61 (1927).

Bettmann, M. A., Paulin, S.: The incidence, nature and modification of undesirable side effects. Radiology *122*, 101 (1977).

Bjordal, R. J.: Simultaneous pressure and flow recordings in varicose veins of the lower extremity. Acta Chir. Scand. *136*, 309 (1970).

Bjordal, R. J.: Circulation patterns in incompetent perforating veins in the calf and in the saphenous system in primary varicose veins. Acta Chir. Scand. *138*, 251 (1972).

Bjordal, R. J.: Blood circulation in varicose veins of the lower extremities Aniology *23*, 163 (1972 a).

Bjordal, R.: Die Zirkulation in insuffizienten V. perforantes der Wade bei venösen Störungen. In: Venae perforantes (May, R., Partsch, H., Straubesand, J., eds.). München-Wien-Baltimore: Urban & Schwarzenberg. 1981.

Bjordal, R. J.: Klinische und therapeutische Konsequenzen der beobachteten hämodynamischen Strömungsmuster in Venae perforantes. In: Venae perforantes (May, R., Partsch, H., Staubesand, J., eds.). München-Wien-Baltimore: Urban & Schwarzenberg. 1981 a.

Bjordal, R. J.: Dilated perforating veins – a dilemma? Congress, Societas Phlebologica Scandinavica, Island of Ven, Sweden, May 5–7, 1982.

Blazek, V.: Medizinisch-technische Grundlagen der Lichtreflexions-Rheographie. In: Die Licht-Reflexions-Rheographie (May, R., Stemmer, R., eds.). Erlangen: Perimed. 1984.

Böck, P.: Vergleich der Morphologie von Arterien- und Venenklappen. Verh. Anat. Ges. *69*, 145 (1975).

Bolliger, A.: Intravasale Fließgeschwindigkeitsmessungen mit dem Laufzeit-Ultraschall-Strömungsmesser in großen insuffizienten Venae perforantes am Unterschenkel. In: Venae perforantes (May, R., Partsch, H., Staubesand, J., eds.) München-Wien-Baltimore: Urban & Schwarzenberg. 1981.

Bollinger, A.: Klappenagnesie und -dysplasie der Beinvenen. Schweiz. med. Wschr. *101*, 1348 (1971).

Bollinger, A.: Funktionelle Angiologie. Stuttgart: G. Thieme. 1979.

Braune, W.: Die Oberschenkelvene des Menschen. Leipzig: Veit u. Co. 1873.

Breslau, W., De Weese, J. A.: Successful endophlebectomy of autogenous venous bypass graft. Ann. Surg. *162*, 251 (1965).

Brodny, M. L., Robins, S., Hershman, H. A., De Nuccio, A.: Epididymography, varicocelography and testicular angiography. Their uses in the study of infertility of the male. Fertil. Steril. *6*, 158 (1955).

Burkitt, D. P.: Varicose veins, deep vein thrombosis and haemorroids: Epidemiology and suggested aetiology. Brit. med. J. *2*, 556 (1972).

Bush, H. L., McCabe, M. E., Nabseth, D. C.: Functional injury of vein graft endothelium. Role of hypothermia and distention. Arch. Surg. *119*, 770 (1984).

Bush, H. L.: In discussion to Kroener and Bernstein. Arch. Surg. *116*, 1472 (1981).

Buxton, R. W., Farris, J. M., Moyer, C. A., Coller, F. A.: Surgical treatment of long standing deep phlebitis of the leg. Surgery *15*, 749 (1944).

Bylsma, P.: Graft patency evaluation in CABG: clips versus no-clips: Joint Meeting of the European Society of Cardiovascular and Interventional Radiology and the European College of Angiography, Vienna, April 22–27, 1985.

Callum, K. G., Gray, L. J., Thomas, M. L.: An evaluation of the fluorescein test and phlebography in the detection of incompetent perforating veins. Brit. J. Surg. *60*, 699 (1973).

Carino, M., McGraw, J. Y., Luke, J. C.: Autogenous vein graft replacement of thrombosed deep veins. Surgery *53*, 123 (1964).

Chilvers, A. S., Thomas, H.: Method for the localization of incompetent ankle perforating veins. Brit. med. J. *2*, 577 (1970).

Cockett, F. B., Dodd, H.: The Pathology and Surgery of the Veins of the Lower Limb. London: Churchill Livingstone. 1976.

Comhaire, F., Montagne, R., Kunnen, R.: The value of scrotal thermography as compared with selective retrograde venography of the internal spermatic vein for the diagnosis of "subclinical" varicocele. Fertil. Steril. *27*, 694 (1976).

Conolly, J. E., Harris, E. J., Mills, W.: Autogenous in situ saphenous vein for bypass of femoral-popliteal obliterative disease. Surgery *55*, 144 (1964).

Corson, J. D., Karmody, A. M., Shah, D. M., Leather, R. P.: Retrograde valve incision for insitu vein-arterial bypass utilising a valvulotome. Ann. Royal Coll. Surg. Engl. *66,* 173 (1984).

Cotton, L. T.: Varicose veins gross anatomy and development. Brit. J. Surg. *48,* 589 (1961).

Dale, W. A.: Surgery for chronic peripheral venous hypertension. Contemp. Surg. *16,* 28 (1980).

De Camp, P. I.: In discussion to Raju. Ann. Surg. *197,* 697 (1983).

De l'Aulnait, H.: Recherches anatomiques et physiologiques sur les valvules des veins. Paris: 1854.

De Weese, J. A., Niguidula, F.: The replacement of short segments of veins with functional autogenous vein grafts. Surg. Gyn. Obstet. *110,* 303 (1960).

Diener, L.: Intraoveous phlebography of the lower limb. Postmortem investigation of thrombotic venous disease. Acta Radiol. Suppl. 304 (1971).

Diethrich, E. B.: Significance of venous valves in iliofemoral thrombectomy. Surg. Gyn. Obstet *124,* 1322 (1967).

Dow, J. D.: The venographic localisation of incompetent communicating veins in the leg. Brit. J. Radiol. *24,* 182 (1951).

Downs, A. R., Morrow J. M.: Valvular stenoses in vein grafts. Curr. Topics Surg. Res. *1,* 499 (1969).

Downs, A. R.: Repair of late vein graft occlusions. Arch. Surg. *103,* 639 (1971).

Dubin, L., Amelar, R. D.: Etiologic factors in 1291 consecutive cases of male infertility. Fertil. Steril. *22,* 469 (1971).

Dubin, L., Hotchkiss, R. S.: Testis biopsy in subfertile men with varicocele. Fertil. Steril. *20,* 50 (1969).

Dzillas, P.: Über das Vorkommen von Klappen in kleinsten Venen beim Menschen. Z. Anat. Entwicklungsgesch. *114,* 309 (1949).

Edwards, A.: The treatment of varicose veins; anatomical factors of ligation of the great saphenous vein. Surg. Gyn. Obstet. *59,* 916 (1934).

Edwards, E. A.: The orientation of venous valves in relation to body surfaces. Anat. Rec. *64,* 369 (1936).

Edwards, A. E., Edwards, J. E.: The effect of thrombophlebitis on the venous valve. Surg. Gyn. Obstet. *65,* 310 (1937).

Edwards, J. E., Edwards, E. A.: The saphenous valves in varicose veins. Americ. Heart J. *19,* 338 (1940).

Eger, S. A., Casper, S. L.: Etiology of Varicose veins from an anatomic aspect, based on dissection of 38 adult cadavers. J. amer. med. Assoc. *123,* 148 (1943).

Eger, S. A., Wagner, F. B.: Etiology of varicose veins. Postgrad. Med. *6,* 243 (1949).

Ehringer, H.: Die Plethysmographie. In: Venöse Abflußstörungen (Ehringer, H., Fischer, H., Netzer, C. O., Schmutzler, R., Zeitler, E., eds.). Stuttgart: F. Enke. 1979.

Eisemann, B., Malette, W.: An operative technique for the construction of venous valves. Surg. Gyn. Obstet. *97,* 731 (1953).

Elem, B., Shorey, A., Lloyd Williams, K.: Comparison between thermography and fluorescin test in the detection of incompetent perforating veins. Brit. med. J. *4*, 651 (1971).

Ellenberger, W., Baum, A.: Handbuch der vergleichenden Anatomie der Haustiere, 12. Ausg., S. 637. 1908.

Eriksson, I., Almgren, B., Norgren, L.: Evaluation of femoral valvuloplasty in primary deep venous insufficiency. VASA *12*, 305 (1983).

Estienne, C.: De dissectione partium corporis humani, pp. 182–357. Paris: 1545.

Fabricius, H. ab Aquapendente. De venarum ostiolis. Padua: 1603.

Fagarasanu, J.: Recherchen anatomiques sur la veine rénale gauche et ses collatérales, leurs rapport avec la pathogénie du varicocele et des varices. Ann. Anat. Path. *15*, 9 (1938).

Fegan, W. G., Kline L.: The course of varicosities in superficial veins of the lower limb. Brit. J. Surg. *59*, 798 (1972).

Fell, St. C.: In discussion to Kistner, R. L. Arch. Surg. *110*, 1341 (1975).

Ferris, E. B., Kistner, R. L.: Femoral vein reconstruction in the management of chronic venous insufficiency, a 14 year experience. Arch. Surg. *117*, 1571 (1982).

Fischer, H.: Verödungstherapie von Varizen. Ärztliche Praxis *24*, 207 (1972).

Fischer, N., Staubesand, J.: Zur Ultrastruktur kollagener Fibrillen in normalen und veränderten Blutgefäßen. Acta anatom. *114*, 125 (1982).

Fogarty, T. J.: In discussion to E. B. Ferris und R. L. Kistner. Arch. Surg. *117*, 1578 (1982).

Folse, R.: The influence of femoral vein dynamics on the development of varicose veins. Surgery *68*, 974 (1970).

Franklin, K. J.: Valves in veins: An historical survey. Proc. Roy. Soc. Med. *21*, 1 (1927).

Franklin, K. J.: Valves in veins, further observations. J. Anat. *64*, 67 (1929).

Friedrich, N.: Über das Verhalten der Klappen in den Cruralvenen sowie über das Vorkommen von Klappen in den großen Venenstämmen des Unterleibes. Morph. Jahrbuch *7*, 323 (1882).

Gasser, G.: Varikozele. Wien. klin. Wschr. *83*, 484 (1971).

Gasser, G., Strasser, R., Pokieser, H.: Thermogramm des Hodens und Spermiogramm. Andrologia *5*, 127 (1973).

Gaudini, V. A., Miller, D. C., Kosek, J. C.: Veno-venous allograft, patency, subendothelial proliferation and the role of platelet active agents. J. Surg. Res. *34*, 263 (1983).

Gibbs, N. M.: Venous thrombosis of the lower limbs with particular reference to bed rest. Brit. J. Surg. *45*, 209 (1957).

Gottlob, R.: Über die Vorbeugung von Schäden bei der Varizenoperation. Prophylaxe *7*, 56 (1968).

Gottlob, R.: Die primäre Varikose, nahezu immer eine Stammvarikose mit geringer Bedeutung der Rami perforantes. In: Ergebnisse der Angiologie, Vol. 19. XIX. Jahreskongreß der Deutschen Gesellschaft für Phlebologie und Protokologie gemeinsam mit der Schweizerischen Gesellschaft für

Phlebologie Wien, 18.–20. November 1977 (Santler, R., Lindemayr, H., Bollinger, A., eds.). Stuttgart-New York: Schattauer. 1978.

Gottlob, R.: The preservation of venous endothelium by "dissection without touching" and by an atraumatic technique of vascular anastomosis. Minerva chirurgica *32*, 693 (1977 a).

Gottlob, R.: Zur Pathogenese und Therapie des Krampfaderleidens. Wien. klin. Wschr. *90*, 1 (1978).

Gottlob, R.: Tierexperimentelle Studie zu Gefaßwändschäden als Ursache früher Venentransplantatthrombosen. Z. exper. Chir. *11*, 43 (1978 a).

Gottlob, R.: Untersuchungen über die endothelschädigende Wirkung von Metrizamide im Vergleich zu einem Diatrizoat und einem Iotalamat. In: Amipaque Workshop (Frommhold, W., Hacker, H., Schmitt, H. E., Vogelsang, H., eds.). Amsterdam-Oxford: Excerpta Medica. 1978.

Gottlob, R.: Lokale Kontrastmittelschäden, Ursachen, Testmethoden, Ergebnisse – in Zukunft der Angiographie, Schmerzfreiheit? Konstanz: Byk Gulden. 1979.

Gottlob, R.: Die klinische Bedeutungslosigkeit der Rami perforantes oder communicantes bei der primären Varikose. In: Venae perforantes (May, R., Partsch, H., Staubesand, J., eds.). München-Wien-Balimore: Urban & Schwarzenberg. 1980.

Gottlob, R.: Intraoperative changes in veins used for arterial replacement with special reference to endothelial injuries and their repair. In: Serono Symposium Nr. 44: Peripheral Arterial Diseases, Medical and Surgical Problems (Stipa, S., Cavellaro, A., eds.), p. 187. London-New York: Academic Press. 1982.

Gottlob, R., Donas, P., El Nashef, B.: Untersuchungen an den obersten Klappen der Vena saphena magna bei Varizen. Zbl. Chir. *100*, 1306 (1975).

Gottlob, R., Donas, P., El Nashef, B.: Untersuchungen am Endothel arterialisierter Venen. I. Die Wirkung einer „atraumatischen Präparation". VASA *4*, 243 (1975 a).

Gottlob, R., Donas, P., El Nashef, B., Saghir, F.: Untersuchungen am Endothel arterialisierter Venen. II. Morphologische Befunde bei autologen Venenimplantaten von Hunden. VASA 5, 111 (1976).

Gottlob, R., Donas, P., El Nashef, B.: Untersuchungen am Endothel arterialisierter Venen. III. Unterschiede zwischen autologen und homologen Transplantaten. VASA *5*, 313 (1976 a).

Gottlob, R., Hoff, H. F.: A study of the relation between endothelial silver lines, medial transverse silver lines and the ultrastructural morphology of blood vessels. Vasc. Surg. *1*, 92 (1967).

Gottlob, R., Hoff, H. F.: Histochemical investigations on the nature of large blood vessels endothelial and medial argyrophilic lines and on the mechanism of silver staining. Histochemie *13*, 70 (1968).

Gottlob, R., Kimmel, A.: Postthrombotische Veränderungen an Venenklappen im Tierexperiment. Virchows Arch. path. Anat. *334*, 337 (1973).

Gottlob, R., May, R.: Der Frühverschluß der Palma-Operation, Vorbeugung durch Anwendung einer endothelschonenden Operationstechnik. VASA 6, 263 (1977).

Gottlob, R., Piza, H., Gottlob, I., Gestring, G. F.: Replacement of small veins by autologous grafts: Application of an endothelium preserving technique. Vasc. Surg. 16, 27 (1982).

Gottlob, R., Saghir, F.: Polypen und Korbhenkelartige Wülste im Klappenbereich variköser Venen. Phlebol. u. Proktol. 5, 1 (1976).

Gottlob, R., Stockinger, L., Gestring, G. F.: Conservation of veins with preservation of viable endothelia. J. Cardiovasc. Surg. 23, 109 (1982 b).

Gottlob, R., Zinner, G.: Über die Regeneration von Endothelien nach hartem und weichem Trauma. Virch. Arch. path. Anat. 336, 16 (1962).

Gullmo, A.: Periphere Venen. In: Handbuch der medizinischen Radiologie, Bd. 10, 3. Teil. Berlin-Göttingen-Heidelberg-New York: Springer. 1964.

Hach, W.,: Spezielle Diagnostik der primären Varikose. Gräfelfing: Demeter. 1981.

Hach, W., Salzmann, G.: Die Chirurgie der Venen. Stuttgart-New York: Schattauer. 1982.

Haimovici, H., Hoffer, P. W., Zincola, N.: An experimental and clinical evaluation of grafts in the venous system. Surg. Gyn. Obstet. 131, 1173 (1970).

Hall, K. V.: The great saphenous vein used in situ as an arterial shunt after extirpation of the vein valves. Surgery 51, 492 (1962).

Hallberg, D.: A method for repairing incompetent valves in deep veins. Acta Chir. Scand. 138, 143 (1972).

Haller, A.: Elementa Physiologica. Corporis humani I/136. 1757.

Haller, J. A., jr.: Thrombectomy for acute iliofemoral thrombosis. Arch. Surg. 83, 448 (1961).

Halstuk, K., Mahler, D., Baker, W. H.: Late sequelae of deep venous thrombosis. Amer. J. Surg. 147, 216 (1984).

Hamer, J. D., Malone, P. O., Silver, J. A.: The pO_2 in venous valve pockets: its possible bearing on thrombogenesis. Brit. J. Surg. 68, 166 (1981).

Harjola, P. T.: Fibrotic valvular stenosis in saphenous vein grafts. Extr. de la Rev. Angéiologie 4, 147 (1982).

Harvey, W.: Exercitatio anatomica de motu cordis et sanguinis in animalibus. 1628.

Hasebrock, K.: Über die Bedeutung der Arterienpulsation für die Strömung in denen Venen und die Pathogenese der Varizen. Pflügers Arch. ges. Physiol. 163, 191 (1916).

Havig, O.: Deep vein thrombosis and pulmonary embolism. Acta Chir. Scand. Suppl. 478 (1977).

Hepp, W.: Zur kongenitalen Aplasie und Avalvulie der Beinvenen. VASA 9, 316 (1980).

Hoare, M. C., Nicolaides, A. N., Miles, C. R., Shull, K., Jury, R. P., Needham, T., Dudley, H. A. F.: The role of primary varicose veins in venous ulceration. Surgery 92, 450 (1982).

Hochstetter, F.: Über das normale Vorkommen von Klappen in den Magenverzweigungen der Pfortader beim Menschen und bei einigen Säugetieren. Arch. Anat. Physiol. 137 (1987).

Hoff, H. F., Gottlob, R.: Ultrastructural changes of large rabbit blood vessels following mild mechanical trauma. Virch. Arch. path. Anat. *345,* 93 (1968).

Hogensgard, I. C.: Phlebography in chronic venous insufficiency of the lower extremity. Acta Radiol. *32,* 375 (1943).

Hornaus, J.: The operative treatment of varicose veins and ulcers, based upon a classification of the lesion. Surg. Gyn. Obstet. *22,* 143 (1916).

Huse, J. B., Nabseth, A. C., Bush, H. L., jr., Widrich, W. C., Johnson, W. C.: Direct venous surgery for venous valvular insufficiency of the lower extremity. Arch. Surg. *118,* 717 (1983).

Husni, E. A.: In situ sapheno-popliteal bypass graft for incompetence of the femoral and popliteal veins. Surg. Gyn. Obstet. *130,* 279 (1970).

Hyde, G. L., Hull, D. A.: Longterm results of subfascial vein ligation for venous stasis disease. Surg. Gyn. Obstet. *153,* 683 (1981).

Jacobson, J. H., jr., Katsumura, T.: Small vein reconstruction. J. Cardiovasc. Surg. *6,* 157 (1965).

Jäger, O.: Die Entwicklung der Venenklappen. Morph. Jahrbuch *56,* 250 (1926).

Johnson, N. D., Queral, L. A., Flynn, W. R., Yao, J. S. T., Bergan, J. J.: Late objective assessment of venous valve surgery. Arch. Surg. *116,* 1461 (1981).

Kampmeier, O. F., Birch, L. F.: The origin and development of the venous valves with particular reference to the saphenous district. Amer. J. Anat. *38,* 451 (1917).

Kampmeier, O. F., Caroll, W. W.: The origin and development of the venous valves, with particular reference to the saphenous district. Amer. J. Anat. *38,* 451 (1928).

Kantrowitz, A., Lerrick, A.: The augmentation of peripheral arterial blood flow by the use of a valve. Surg. Forum *3,* 151 (1952).

Karino, T., Motomiya, M.: Flow patterns in the pockets of venous valves and their implication in thrombogenesis. Abstract, International Symposium on Venous Diseases of the lower limbs, Florenz, 10. – 12. März 1982.

Karino, T., Motomiya, M.: Flow through a venous valve and its implication for thrombus formation. Thrombosis Research. *1984.*

Keith, A.: An account of the structure concerned in the production of the jugular pulse. J. amer. Physiol. *42,* 1 (1907).

Kempczinski, R. F.: In discussion to N. D. Johnson et. al. Arch. Surg. *116,* 1466 (1981).

Kistner, R. L.: Surgical repair of the incompetent femoral vein valve. Arch. Surg. *110,* 1336 (1975).

Kistner, R. L., Sparkuhl, M. D.: Surgery in acute and chronic venous disease. Surgery *85,* 31 (1979).

Klotz, K.: Untersuchungen über die Vena saphena magna beim Menschen, besonders rücksichtlich ihrer Klappenverhältnisse. Arch. Anat. Physiol. Anatom. Abtlg., 159 (1887).

Kojima, Y., Sanada, Y., Fonkalsrud, E. W.: The effect of immunosuppression on the patency and structure of canine vein allografts in the venous system. Surg. Gyn. Obstet. *154*, 853 (1982).

Kormano, M., Kahanpää, K., Svinhufvud, U., Tähti, E.: Thermography of varicocele. Fertil. Steril. *21*, 558 (1970).

Kormano, M., Kahanpää, K., Tähti, E.: Thermographic recording of the reaction of normal and varicocele scrotum to increased temperature. Andrologia *5*, 201 (1973).

Kriessmann, A., Bollinger, A., Keller, H.: Praxis der Doppler-Sonographie. Stuttgart: G. Thieme. 1982.

Kroener, J. M., Bernstein, E. F.: Valve competence following experimental valve transplantation. Arch. Surg. *116*, 1467 (1981).

Krupski, W., Thal, E. R., Gewertz, B. L., Buja, M., Murphy, M. E., Hagler, H. K., Frey, W. J.: Endothelial response to venous injury. Arch. Surg. *114*, 1240 (1979).

Kubik, St., May, R.: Das Verspannungssystem der Venen in der Subinguinalregion. In: Die Leiste. Aktuelle Probleme in der Angiologie, Vol. 38 (Brunner, U., ed.). Bern-Stuttgart-Wien: H. Huber. 1979.

Kunlin, J. (1953): Les greffes veineuses. 15me Congrès de la Société Internationale de Chirurgie, Lissabon, September 14–20, 1953 (Société Internationale de Chirurgie, eds.).

Kunlin, J.: Experimentelle Venenchirurgie. In: Chirurgie der Bein- und Beckenvenen (May, R., ed.). Stuttgart: G. Thieme. 1974.

Kügelgen, A. v., Greinemann, H.: Die Klappen in den menschlichen Nierenvenen, besonders in der Mündung der Nierenbeckenvenen. Z. Zellforsch. *47*, 648 (1958).

Laerum, F., Holm, H. A.: Postphlebographic thrombosis. Radiology *140*, 651 (1981).

Laerum, F., Börsum, T., Reisaag, A.: Human endothelial cell culture as an evaluation system for the toxicity of intravascular contrast media. Invest. Radiol. *18*, 199 (1983).

Langeron, P., Puppinck, P., Gordonnier, D.: La technique de la greffe veineuse "in situ" dans la chirurgie artérielle restauratrice des membres inférieures. J. de chir. *115*, 171 (1978).

Lanz, T. v., Wachsmuth, W.: Praktische Anatomie. Bein und Statik. Berlin: Springer. 1938.

Leather, R. P., Schah, D. M., Karmody, A. M.: Infrapopliteal arterial bypass for limb salvage: Increased patency and utilization of the saphenous vein "in situ". Surgery *90*, 1000 (1981).

Lea Thomas, M.: Gangrene following periperal phlebography of the legs. Brit. J. Radiol. *43*, 528 (1970).

Lea Thomas, M.: Phlebography. In: Complications in Diagnostic Radiology (Ansell, G., ed.). Philadelphia: Lippincott. 1976.

Lea Thomas, M., Macdonald, L. M.: Complications of ascending phlebography of the leg. Brit. med. J. *2*, 317 (1978).

Ledderhose, G.: Studien über den Blutkreislauf in den Hautvenen unter physiologischen und pathologischen Bedingungen. Mitt. Grenzgeb. Med. Chir. *14* (1905).

Lee, B. Y., Kavner, D., Thoden, W. R., Trainor, F. S, Lewis, J. M., Malden, J. L.: Technique of venous impedance plethysmography for quantification of venous reflux. Surg. Gyn. Obstet. *154*, 49 (1982).

Le Page, J. R.: Gangrene of the foot: A complication of phlebography. Angiology *36*, 399 (1885).

Leu, H. J.: Die moderne meßtechnische Beurteilung venöser Erkrankungen. Med. Welt *20*, (N. F.) 2242 (1969).

Leu, H. J.: Histopathologie der Venenerkrankungen. Bern-Stuttgart-Wien: H. Huber. 1971.

Leu, H. J., Vogt, M., Pfrundes, H.: Morphological alterations of non-varicose and varicose veins. A morphological contribution to the discussion on pathogenesis of varicose veins. Basic Res. Cardiol. *74*, 435 (1979).

Lima, S. S., Castro, M. P., Costa, O. F.: A new method for the treatment of varicocele. Andrologia *10*, 103 (1978).

Lindvall, N., Lodin, A.: Congenital absence of valves in the deep veins of the leg. Acta Dermato-Venerologica Scand. *41*, Suppl. 45 (1961).

Linton, R. R.: The communicating veins of the lower leg and the operative technique for their ligation. Ann. Surg. *107*, 582 (1938).

Linton, R. R., Hardy, F. B.: Postthrombotic Syndrome of the lower extremity, treated by interruption of the superficial femoral vein and ligation and stripping of the long and short saphenous vein. Surgery *24*, 452 (1948).

Ljungnér, H., Bergquist, D.: Decreased fibrinolytic activity in the bottom of human vein valve pockets. VASA *12*, 333 (1983).

Lodin, A., Zetterquist, St.: Metabolic evaluation of the crural blood flow during standing and during sitting leg exercise in patients with congenital absence of venous valves in the legs. Acta derm.-vener. *45*, 26 (1965).

Lofgren, E. P., Myers, T. T., Lofgren, K. A., Kuster, K.: The venous valves of the foot and ankle. Surg. Gyn. Obstet. *127*, 289 (1968).

Ludbrook, J.: Aspects of venous function in the lower limbs. Springfield, Ill.: Ch. C Thomas. 1968.

Ludbrook, J.: The analysis of the venous system. Bern: H. Huber. 1972.

Ludbrook, J., Beale, G.: Femoral venous valves in relation to varicose veins. Lancet *i*, 79 (1962).

Luft, J. H.: Fine structure of capillary and endocapillary layers as revealed by ruthenium red. Feder. Proc. *25*, 1773 (1966).

Luke, J. C.: Retrograde venography of the deep veins. J. canad. med. Assoc. *49*, 86 (1943).

Luke, J. C.: The deep vein valves. A venographic study in normal and postphlebitic states. Surgery *29*, 381 (1951).

Luke, J. C.: The management of recurrent varicose veins. Surgery *35*, 40 (1954).

Lund, F. L., Diener, L., Ericsson, J. L. E.: Postmortem intraosseous phlebography as an aid in studies of venous thromboembolism. Angiology *20*, 155 (1969).

Lusitanus Amatus. Curationum medicinalium, centuria prima, scholia curationis 52. 1551.

Malette, W. G.: In: Discussion to N. D. Johnson et al. Arch. Surg. *116*, 1465 (1981).

Mali, W., Oei, H., Coolsaet, B., Schnur, K., Arndt, J., Kramer, J.: Haemodynamic aspects of the varicocele. Joint Meeting of the European Society of Cardiovascular and Interventional Radiology and the European College of Angiography, Vienna, April 22–27, 1985.

Maros, T.: Data regarding the typology and the functional significance of the venous valves. Morphology and Embryology *27*, 195 (1981).

Marsman, J. W. P.: Sufficient valves in the spermatic veins: still varicocele. Joint Meeting of the European Society of Cardiovascular and Interventional Radiology and the European College of Angiography, Vienna, April 22–27, 1985.

Martinet, J. D.: In: Die Phlebographie der unteren Extremität (May, R., Nissl, R., ed.). Stuttgart: G. Thieme. 1959.

Mason, R., Giron, F.: Noninvasive evaluation of venous function in chronic venous disorders. Surgery *91*, 312 (1982).

May, R.: Venentransplantationen beim postthrombotischen Zustandsbild. Actuelle Chirurgie *7*, 1 (1972).

May, R.: Venenplastiken bei postthrombotischen Zustandsbildern. Acta Chirurg. Austriaca Suppl. *4*, 1 (1972a).

May, R.: Surgery of the Veins of the Leg and Pelvis. Stuttgart: G. Thieme. 1979.

May, R.: Spätergebnisse nach Femoralisbypass. VASA *8*, 67 (1979a).

May, R.: Die Chirurgie der Varizen der inneren und äußeren Knöchelgegend. In: Der Fuß (Brunner, U., Hrsg.). Bern: H. Huber. 1982.

May, R., Kriessmann, A.: Die periphere Venendruckmessung. Stuttgart: G. Thieme. 1978.

May, R., Mignon, G.: Thrombophlebitis nach Phlebographie. In: Amipaque-Workshop (Frommhold, W., Hacker, H., Schmitt, H. E., Vogelsang, H., eds.). Amsterdam-Oxford: Excerpta Medica. 1978.

May, R., Nissl, R.: Die Phlebographie der unteren Extremität. Stuttgart: G. Thieme. 1973.

May, R., Partsch, H., Staubesand, J. (eds.): Venae perforantes. München-Wien-Baltimore: Urban & Schwarzenberg. 1981.

May, R., Stemmer, R.: Die Licht-Reflexions-Rheographie. Erlangen: Perimed. 1984.

May, R., Thurner, J.: Ein Gefäß-Sporn in der Vena iliaca sinistra als Ursache der überwiegenden linksseitigen Beckenvenenthrombosen. Z. Kreislaufforsch. *45*, 912 (1956).

McLachlin, A. D., McLachin, J. A., Jory, T. A., Rawling, E. G.: Venous stasis in the lower extremities. Ann. Surg. *152*, 678 (1960).

McLeod, J.: Seminal cytology in the presence of varicocele. Fertil. Steril. *16,* 735 (1965).

McNamara, J. J., Darling, R. C., Linton, R. R.: Segmental stenosis of saphenous vein autografts. New Engl. J. Med. *277,* 290 (1967).

Melville, H. (1851): Moby Dick, p. 344. (New America Library of World Literature.) New York: 1961.

Meriel, P.: Note sur les distances valvulaires dans quelques vènes. Arch. anat. *6,* 99 (1926).

Nachbur, B.: Assessment of venous disorders by venous pressure measurements. In: The Treatment of Venous Disorders (Hobbs, J. T., ed.). Philadelphia: J. B. Lippincott. 1977.

Netzer, C. O.: Venöse Abfluß-Störungen. (Together with H. Ehringer, H. Fischer, R. Schmutzler, and E. Zeitler.) Stuttgart: Enke. 1979.

Noble, J., Gunn, A. A.: Varicose veins: comparative study of methods for detecting incompetent perforators. Lancet 7763, 1253 (1972).

Norgren, L.: Functional evaluation of chronic venous insufficiency. Acta Chir. Scand. Suppl. 444 (1974).

Nyboer, J.: Electrical Impedance Plethysmography. Springfield, Ill.: Ch. C Thomas. 1970.

Nyman, U., Almén, T.: Effects of contrast media on aortic Endothelium. Acta Radiol. Suppl. 362, 65 (1980).

O'Neill, J. F.: The effects on venous endothelium of alterations in the blood flow through the vessels in vein walls and the possible relation to thrombosis. Ann. Surg. *126,* 270 (1947).

Otto, K.: Die Problematik der klinischen und röntgenologischen Diagnostik der Beinvenenerkrankungen. Zbl. Phlebologie *2,* 115 (1963).

Palma, E. C.: Vein transplants and grafts in the surgical treatment of the postphlebitic syndrome. J. Cardiovasc. Surg. *1,* 94 (1960).

Palomo, A.: Radical cure of varicocele by a new technique, preliminary report. J. Urol. *61,* 604 (1949).

Paterson, J. C., McLachlin, J.: Precipitating factors in venous thrombosois. Surg. Gyn. Obstet. *98,* 96 (1954).

Patil, K. D., Williams, J. R., Lloyd Williams, K.: Thermographic localization of incompetent perforating veins in the leg. Brit. med. J. *1,* 195 (1970).

Perthes, G.: Über die Operation der Unterschenkelvarizen nach Trendelenburg. Dtsch. med. Wschr. *21,* 253 (1895).

Piulachs, P., Vidal Barraquer, F.: Pathogenetic study of varicose veins. Angiology *4,* 59 (1953).

Plate, G., Brudin, L., Eklöf, B., Jensen, R., Ohlin, P.: Physiologic and therapeutic aspects in congenital vein valve aplasia of the lower limb. Ann. Surg. *198,* 229 (1983).

Ponomarenko, V. N.: Structural features, biochemical properties of the cusps of venous walls of lower extremity veins in humans Arkh. Anat. Gistol. Embriol. *78,* 36 (1980). (Abstr.: Index Medicus 8010.)

Powell, T., Lynn, R. B.: The valves of the external iliac, femoral and upper third of the popliteal veins. Surg. Gyn. Obstet. *92,* 453 (1951).

Psathakis, N.: Ein neues Operationsverfahren zur rationellen Behandlung des Insuffizienz-Syndroms der tiefen Beinvenen. Der Chirurg *35*, 79 (1964).

Psathakis, N.: Has the "substitute valve" at the popliteal vein solved the problem of venous insufficiency of the lower extremities? J. Cardiovasc. Surg. *9*, 64 (1968).

Queral. L. A., Neiman, H. L., Yao, J. St., Bergan, J. J.: Surgical correction of chronic deep venous insufficiency by valvular transplantation. Surgery *87*, 688 (1980).

Raivio, E.: Untersuchungen über die Venen der unteren Extremitäten mit besonderer Berücksichtigung der gegenseitigen Verbindungen zwischen den oberflächlichen und tiefen Venen. Ann. Med. Exp. Fenn. *26*, Suppl. 4 (1948).

Raju, S.: In: Discussion to N. D. Johnson et al. Arch. Surg. *116*, 1466 (1981).

Raju, S.: In: Discussion to E. B. Ferris and R. L. Kistner. Arch. Surg. *117*, 1579 (1982).

Raju, S.: Venous insufficiency of the lower limb and stasis ulceration. Ann. Surg. *197*, 688 (1983).

Raju, S., Perry, J. T.: The response of venous valvular endothelium to autotransplantation and in vivo preservation. Surgery *94*, 770 (1983).

Reagan, B., Folse, R.: Lower limb venous dynamics in normal persons and children of patients with varicose veins. Surg. Gyn. Obstet. *132*, 15 (1971).

Recek, C.: A critical appraisal of the rôle of ankle perforators for the genesis of venous ulcers in the lower leg. J. Cardiovasc. Surg. *12*, 45 (1971).

Recek, C., Koudelka, V.: Effet circulatoire du reflux saphènien dans la varicose essentielle. Phlébologie *32*, 407 (1979).

Rickenbacher, J.: Zur Entwicklung der Venen der unteren Extremität. Zbl. Phlebologie *5*, 6 (1966).

Riedl, P.: Selektive Phlebographie und Katheterthrombosierung der Vena testicularis bei primärer Varicocele. Wien. klin. Wschr. *91*, Suppl. 99 (1979).

Robb, W. A. T.: Operative treatment of varicocele. Brit. med. J. *2*, 355 (1955).

Rushton, F.: In: Discussion to E. B. Ferris and R. L. Kistner. Arch. Surg. *117*, 1579 (1982).

Salomonowitz, E., Gottlob, R.: Untersuchungen am Endothel der Vena iliaca communis sinistra als Beitrag zur Pathogenese des Venensporna. VASA *10*, 194 (1981).

Samuels, P. B., Plested, W. G., Haberfelde, G. C., Cincotti, J. J., Brown, C. E.: In situ saphenous veins arterial bypass: A study of the anatomy pertinent to its use in situ as a bypass graft with a description of a new valvulotome. Amer. Surg. *34*, 122 (1968).

Sandmann, W. C., Nüllen, H., Ditmer, J., Landwehr, B.: Chirurgische Therapie der primären Femoralklappeninsuffizienz. In: Chirurgie der Venen (Hach, W., Salzmann, G., eds.). Stuttgart-New York: Schattauer. 1982.

Santler, R.: Kritik zur Arbeit „Führen Krampfadern ihren Namen zu Recht?" VASA *3*, 210 (1974).

Saphir, O., Lev, M.: Venous valvulitis. Arch. Path. *53*, 456 (1952a).

Saphir, O., Lev, M.: The venous valve in the aged. Amer. Heart J. *44*, 843 (1952 a).

Schaeffer, H.: Die Anonyma-Punktion und ihre Beziehung zu den venösen Grenzklappen. Münch. med. Wschr. *115*, 782 (1973).

Schanzer, H. C.: In: Discussion to E. B. Ferris and R. L. Kistner. Arch. Surg. *117*, 1578 (1982).

Schanzer, H., Lande, L., Premur, E., Peirce, C.: Noninvasive evaluation of chronic venous insufficiency. Use of foot mercury strain-gauge plethysmography. Arch. Surg. *119*, 1013 (1984).

Schlunk, T., Göltner, E.; Doppler-Prüfung der Venenklappen-Funktion und Varikosis. Münch. med. Wschr. *115*, 782 (1973).

Sevitt, S.: Organization of valve pocket thrombi and the anomalies of double thrombi and valve cusp involvement. Brit. J. Surg. *61*, 641 (1974).

Sevitt, A.: The structure and growth of valve-pocket thrombi in femoral veins. J. Clin. Path. *27*, 517 (1974 a).

Shankin, W. M., Axam, N. A.: On the presence of valves in the rats cerebral arteries. Anat. Rec. *146*, 145 (1963).

Sheppard, M.: A procedure for the prevention of saphenofemoral incompetence. Austr.-New Zealand J. Surg. *48*, 322 (1978).

Sherman, R. S.: Varicose veins. Further findings based on anatomic and surgical dissections. Ann. Surg. *130*, 218 (1949).

Shin, C. S., Hatem, J. N., Abaci, I. F.: Effect of diminished distal blood flow on the morphologic changes in autogenous vein grafts. Surg. Gyn. Obstet. *147*, 189 (1978).

Shull, K. C., Nicolaides, A. N., Fernandes, J. F., Miles, C., Horner, J., Needham, T., Cooke, E. D., Eastcott, F. H. G.: Significance of the popliteal reflux in relation to ambulatory venous pressure and ulceration. Arch. Surg. *114*, 1304 (1979).

Spiess, H. F.: Ultraschall-Doppler-Sonde bei Erkrankungen im Venensystem. Die Kapsel *29*, 1253 (1972).

Staubesand, J.: Kleiner Atlas zur vaskulären Anatomie der Leistengegend. In: Die Leiste (Brunner, U., ed.). Bern: H. Huber. 1979.

Staubesand, J., Rulffs, W.: Die Klappen kleiner Venen. Z. Anat. Entwicklungsgesch. *120*, 392 (1958).

Suchard, M. E.: Observations nouvelles sur la stricture du tronc de la veine porte du rat, du lapin, du chien de l'homme et du poulet. C. R. Soc. Biol. (Paris) 192 (1901).

Sylvius, J.: Isagoge Anatomica. 1555.

Szilagyi, D. E., Elliot, J. P., Hageman, J. H., Smith, R. F., Dall-Olmo, C. A.: Biologic fate of autogenous vein implants as arterial substitutes. Ann. Surg. *178*, 232 (1973).

Szotér, T., Cronin, R. F. P.: Venous distensibility in patients with varicose veins. Canad. med. Assoc. J. *94*, 1293 (1966).

Taheri, S. A., Heffner, R., Williams, J., Lazar, L., Elias, St.: Muscle changes in venous insufficiency. Arch. Surg. *119*, 929 (1984).

Taheri, S. A., Lazar, L., Elias, St.: Status of vein valve transplantation after 12 months. Arch. Surg. *117*, 1313 (1982).

Thulesius, O., Norgren, L., Gjöres, J. C.: Foot volumetry a new method for objective assessment of edema and venous function. VASA *2*, 325 (1973).

Tibbs, D. J., Fletcher, E. W. L.: Direction of flow in superficial veins as a guide to venous disorders of the lower limb. Surgery *93*, 758 (1983).

Todd, A. S.: The histological localization of fibrinolysinactivator. J. Path. Bact. *78*, 281 (1959).

Tompa, G., Gyöngyössi, G., Géhl, A., Arday, G.: Der Ersatz von Venenklappen im Tierversuch. Z. exper. Med. *8*, 148 (1975).

Trendelenburg, F.: Über die Unterbindung der Vena saphena magna bei Unterschenkelvarizen. Bruns Beitr. Klin. Chir. *7*, 195 (1890).

Tsapagos, M. J.: In: Discussion to J. M. Kroener and E. F. Bernstein. Arch. Surg. *116*, 1471 (1981).

Tulloch, U. S.: Varicocele in subfertility: Results and treatment. Brit. med. J. *2*, 356 (1955).

Vollmar, J. F., Hutschenreiter, S.: Temporary arteriovenous fistula. In: Pelvic and Abdominal Veins, International Congress Series, 550 (May, R., Weber, J. eds.). Amsterdam-Oxford-Princeton: Excerpta Medica. 1981.

Waddell, W. G., Vogelfanger, I. J., Prudhomme, P., Ram, J. D., Beathie, W. G., Ewing, J. B.: Venous valve transplantation. Arch. Surg. *88*, 5 (1964).

Wagenvoort, C. A.: Die Bedeutung der Arterienwülste für den Blutkreislauf. Acta Anat. *21*, 70 (1954).

Wall, C. A.: In: Discussion to R. L. Kistner and M. D. Sparkuhl. Surgery *85*, 42 (1979).

Warren, R., Thaya, T. R.: Transplantation of the saphenous vein for postphlebitic stasis. Surgery *35*, 867 (1956).

Welsh, P., Parisi, C., Rosas, G., Regetto, R., Paladino, C.: Degenerative changes in autologous veins used as arterial bypass grafts. J. Cardiovasc. Surg. *15*, 700 (1974).

Weissenhofer, W., Schueller, E. E., Worthington, G., Schenk, G.: The fate of venous valves in the arterial tree. J. Surg. Res. *17*, 200 (1974).

Whitney, R. J.: Strain gauge plethysmography. J. Physiol. *121*, 1 (1953).

Wessler, S., Ward, K., Ho, C.: Studies in intravascular caogaulation. J. Clin. Invest. *34*, 647 (1955).

Wienert, V.: Diagnostik der insuffizienten Perforansvenen – Wertigkeit der Methoden. In: Venae perforantes (May, R., Partsch, H., Staubesand, J., eds.). München-Wien-Baltimore: Urban & Schwarzenberg. 1981.

Williams, A. F.: A comparative study of the venous valves in the limbs. Surg. Gyn. Obstet. *99*, 676 (1954).

Wuppermann, Th., Marées, de H., Reiss, H. D., Knospe, E., Zschegge, Ch.: Zur Beurteilung verschiedener Therapieverfahren bei primärer Varikosis. VASA *7*, 291 (1978).

Yao, J. S. T.: In: Noninvasive Diagnostic Techniques in Vascular Diseases (Bernstein, E., ed.). St. Louis: C. V. Mosby. 1982.

Zinner, G., Gottlob, R.; Morphologic changes in vessel endothelia caused by contrast media. Angiology *10*, 207 (1959).

Subject Index

GPSR Compliance
The European Union's (EU) General Product Safety Regulation (GPSR) is a set
of rules that requires consumer products to be safe and our obligations to
ensure this.

If you have any concerns about our products, you can contact us on

ProductSafety@springernature.com

In case Publisher is established outside the EU, the EU authorized
representative is:

Springer Nature Customer Service Center GmbH
Europaplatz 3
69115 Heidelberg, Germany